I0470708

Energy Storage for Power Grids and Electric Transportation: A Technology Assessment

Paul W. Parfomak
Specialist in Energy and Infrastructure Policy

March 27, 2012

Congressional Research Service

7-5700

www.crs.gov

R42455

Summary

Energy storage technology has great potential to improve electric power grids, to enable growth in renewable electricity generation, and to provide alternatives to oil-derived fuels in the nation's transportation sector. In the electric power system, the promise of this technology lies in its potential to increase grid efficiency and reliability—optimizing power flows and supporting variable power supplies from wind and solar generation. In transportation, vehicles powered by batteries or other electric technologies have the potential to displace vehicles burning gasoline and diesel fuel, reducing associated emissions and demand for oil.

Federal policy makers have become increasingly interested in promoting energy storage technology as a key enabler of broad electric power and transportation sector objectives. The Storage Technology for Renewable and Green Energy Act of 2011 (S. 1845), introduced on November 10, 2011, and the Federal Energy Regulatory Commission's Order 755, *Frequency Regulation Compensation in the Organized Wholesale Power Markets*, are just two recent initiatives intended to promote energy storage deployment in the United States. Numerous private companies and national laboratories, many with federal support, are engaged in storage research and development efforts across a very wide range of technologies and applications.

This report attempts to summarize the current state of knowledge regarding energy storage technologies for both electric power grid and electric vehicle applications. It is intended to serve as a reference for policymakers interested in understanding the range of technologies and applications associated with energy storage, comparing them, when possible, in a structured way to highlight key characteristics relevant to widespread use. While the emphasis is on technology (including key performance metrics such as cost and efficiency), this report also addresses the significant policy, market, and other non-technical factors that may impede storage adoption. It considers eight major categories of storage technology: pumped hydro, compressed air, batteries, capacitors, superconducting magnetic energy storage, flywheels, thermal storage, and hydrogen.

Energy storage technologies for electric applications have achieved various levels of technical and economic maturity in the marketplace. For grid storage, challenges include roundtrip efficiencies that range from under 30% to over 90%. Efficiency losses represent a tradeoff between the increased cost of electricity cycled through storage, and the increased value of greater dispatchability and other services to the grid. The capital cost of many grid storage technologies is also very high relative to conventional alternatives, such as gas-fired power plants, which can be constructed quickly and are perceived as a low risk investment by both regulated utilities and independent power producers. The existing market structures in the electric sector also may undervalue the many services that electricity storage can provide. For transportation storage, the current primary challenges are the limited availability and high costs of both battery-electric and hydrogen-fueled vehicles. Additional challenges are new infrastructure requirements, particularly for hydrogen, which requires new distribution and fueling infrastructure, while battery electric vehicles are limited by range and charging times, especially when compared to conventional gasoline vehicles.

Substantial research and development activities are underway in the United States and elsewhere to improve the economic and technical performance of electricity storage options. Changes to market structures and policies may also be critical components of achieving competitiveness for electricity storage devices. Removing non-technical barriers may be as important as technology improvements in increasing adoption of energy storage to improve grid and vehicle performance.

Contents

Figures

Tables

Appendixes

Contacts

Introduction

Energy storage technology has great potential to improve electric power grids, to enable growth in renewable electricity generation, and to provide alternatives to oil-derived fuels in the nation's transportation sector. In the electric power system, the promise of this technology lies in its potential to increase grid efficiency and reliability—optimizing power flows and supporting variable power supplies from wind and solar generation. In transportation, vehicles powered by batteries or other electric technologies have the potential to displace vehicles burning gasoline and diesel fuel, reducing associated emissions and demand for oil.

In recent years, federal policy makers have become increasingly interested in promoting energy storage technology as a key enabler of broad electric power and transportation sector objectives. In remarks about the STORAGE Act of 2011 (S. 1845),[1] which would provide investment tax credits for storage systems connected to the electric grid, businesses and homes, Senate Energy and Natural Resources Committee Chairman Jeff Bingaman remarked,

> Deployment of storage technologies will make our nation's electricity grid more reliable while also enabling more efficient use of existing energy sources as well as new ones, such as wind and solar.... These technologies have the potential to cut electricity bills, reduce peak power demand and lower greenhouse gas emissions.[2]

Likewise, in a statement regarding new energy storage-related rules for wholesale electricity markets, Federal Energy Regulatory Commissioner John Norris stated,

> I believe today's final rule is a positive first step by the Commission in recognizing the unique characteristics and the value that storage resources offer.... As we move forward, I strongly believe that storage will become ever more critical as we look to integrate increasing amounts of variable energy resources.[3]

Referring to advanced batteries for electric transportation applications, Secretary of Energy Steven Chu reportedly stated,

> It's now within grasp, that you can get a battery where the business plans are one-third of the cost of today's batteries, where you can get ranges now that would allow cars instead of 100 miles on a single charge, go 300 or more miles on the same charge.... It's not a pipe dream 30 years from today or 20 years from today. It's in the next decade.[4]

Statements such as those above highlight not only the technical opportunities for energy storage in the grid and in electric transportation, but also the attention being paid to energy storage technologies at the highest levels in the federal government. Nonetheless, many new energy storage technologies continue to face significant technological and economic challenges to their

---

[1] Storage Technology for Renewable and Green Energy Act of 2011 (S. 1845) introduced on November 10, 2011, by Senator Ron Wyden and co-sponsored by Senators Jeff Bingaman, Susan Collins, and Robert Menendez.

[2] Office of Senator Ron Wyden, "Wyden, Collins, Bingaman Legislation Will Increase Investments in the Storage of Renewable Energy," press release, November 10, 2011.

[3] Commissioner John R. Norris, "Frequency Regulation Compensation in the Organized Wholesale Power Markets," Docket Nos. RM11-7-000 & AD10-11-000, Item No. E-28, Federal Energy Regulatory Commission, October 20, 2011.

[4] Michael Warren, "Energy Secretary Steven Chu on Electric Cars," *The Weekly Standard Blog,* April 3, 2011, http://www.weeklystandard.com/blogs/chu-electric-cars_556135.html.

commercialization and widespread deployment. The recent bankruptcy of Beacon Power, one of the leading developers of flywheel energy storage technologies for the grid, is a prominent illustration of commercial barriers to grid storage technology. Public concerns about elevated fire risks from Chevrolet Volt electric car batteries, although shown to be exaggerated, are another.[5] By contrast, increasing investments by AES Corporation in utility-scale battery storage for power grids show continuing successful efforts to overcome technical challenges and market barriers to bring new storage technologies into the market.[6]

Understanding the potential of energy storage in electric applications is complicated by a number of factors. The first is the wide range of storage technologies either commercially available, in development, or being researched. Because they are technologically diverse, it is difficult to gain a balanced understanding of the fundamental capabilities, costs, and comparative advantages of these different energy storage options. Second, there are multiple applications of energy storage, each with distinct operational requirements. Certain storage technologies may suit certain applications better than others. Finally, there are many aspects of market structure and economic regulation that affect energy storage deployment. Taken together, these factors make the development of an energy storage research and development portfolio challenging. While there is general consensus that storage technology improvements are needed, there are multiple potential pathways to such improvements that cut across different disciplines.

This report attempts to summarize the current state of knowledge regarding energy storage technologies for both electric power grid and electric vehicle applications. It is intended to serve as a reference for policymakers interested in understanding the range of technologies and applications associated with energy storage, comparing them, when possible, in a structured way to highlight key characteristics relevant to widespread use. The report also discusses how aspects of policy and market structure affect competition among both mature and emerging technologies.

Structure of the Report

The report contains 13 chapters, starting with an Executive Summary, which provides an overview of the report's main findings. Context and background are provided in "Chapter 2: Background and Scope" and "Chapter 3: Overview of Storage Technology Applications and Benefits." In particular, Chapter 3 provides an overview of electricity storage applications and value, including their use to enable renewable electricity; current barriers to deployment; and current initiatives to address technical, economic and market barriers. Chapters 4-13 discuss the individual storage technologies. Each chapter can be read independently, but Chapters 2 and 3 offer the reader a more complete understanding of some of the more technologically focused discussions in the subsequent chapters.

---

[5] Jim Henry, "Chevy Volt Battery Fires Threaten All Electric Vehicle Makers, Not Just GM," Forbes, December 12, 2011; National Highway Traffic Safety Administration, "NHTSA Statement on Conclusion of Chevy Volt Investigation," press release, January 20, 2012.

[6] "AES Peaker-Sized Battery Proposals Show Company's Vision of Storage Potential for the Grid," *Electric Utility Week*, Platts, January 2, 2012.

Other CRS Reports on Electricity Storage

CRS has written two previous reports on electricity storage: CRS Report R40797, *Electric Power Storage*, by Stan Mark Kaplan, and CRS Report R41709, *Battery Manufacturing for Hybrid and Electric Vehicles: Policy Issues*, by Bill Canis.

Technology Assessment Authorship

This technology assessment and report was prepared by the National Renewable Energy Laboratory (NREL), Strategic Energy Analysis Center, with contributions from Paul Denholm, Anne Dillon, Easan Drury, Greg Glatzmaier, Jeffrey Logan, Marc Melaina, Jeremy Neubauer, Doug Reindl (University of Wisconsin-Madison), Shriram Santhanagopalan, Kandler Smith, Darlene Steward, and Samir Succar (Natural Resources Defense Council). The work was performed under contract to CRS as part of a multiyear CRS project to examine different aspects of U.S. energy policy. John L. Moore, Assistant Director, Resources, Science, and Industry Division, served as the CRS project coordinator. Paul W. Parfomak, Specialist in Energy and Infrastructure Policy, served as the CRS reviewer and editor of the final report.

Acknowledgement

This report was funded, in part, by a grant from the Joyce Foundation.

Chapter 1: Executive Summary

Background

Energy storage in electric applications can provide two significant benefits to the nation's energy system. First, it can improve the technical and economic performance of the electric power grid, increasing reliability and potentially decreasing costs while allowing greater penetration of intermittent sources like solar and wind generation. Second, it enables a potential transition from an oil-based transportation system to one based on an array of domestically sourced electricity options, greatly reducing dependence on petroleum. In both cases, a reduction in the burning of fossil fuels could result in lower overall U.S. carbon emissions and conventional pollutants.

For purposes of assessment and comparison, it is helpful to organize the various energy storage technologies under two industry sectors (electric grid and transportation) and two general categories of application based on the amount of time the storage device is required to provide service (high power/rapid discharge and energy management), further explained below. **Table 1** lists the storage technologies considered in this report according to these categories. The report provides an overview of the current capabilities and costs of each storage technology, including the potential for technical and cost improvement. It also discusses non-technical barriers including environmental, material, market, and policy challenges to widespread deployment.

Table 1. Energy Storage Applications and Technologies

	Electric Grid (Stationary)	**Transportation (Vehicular)**
High Power / Rapid Discharge	Batteries • Lead-Acid • Nickel • Lithium-Ion Capacitors Flywheels Superconducting Magnetic Energy Storage (SMES)	Batteries • Nickel Capacitors Flywheels
Energy Management	Batteries • Advanced Lead-Acid • Flow • High Temperature Hydrogen Compressed Air Pumped Hydro Thermal • Concentrating Solar Power • End Use	Batteries • Lithium-Ion • Lithium-Metal • Metal Air Hydrogen

Source: P. Denholm, National Renewable Energy Laboratory, 2011.

Note: Electric power and transportation applications may elsewhere be referred to as "stationary" and "vehicular," respectively.

Energy Storage for Electric Grid Applications

It is possible to divide grid storage applications into two broad categories based on the length of time a storage device needs to provide service: (1) high power applications where the device must respond rapidly and be able to discharge for only short-term periods (up to about one hour), and (2) energy management related applications where the device may respond more slowly but must be able to discharge for several hours or more. Ideally, all storage devices would be able to provide all services, but some technologies are technically restricted to provide only short-term services. However, many of these services have very high value in the grid, so short-term storage can still provide considerable benefits.

High Power/Rapid Discharge Applications

The rapid response category can be further divided into short-term discharge—less than one minute—used to provide grid stability and power quality, and longer-term discharge—up to about an hour. Though important, short-term discharge services can often be provided by non-storage options such as power electronics. Furthermore, this class of grid services does not address the primary challenge of renewables integration, which requires minutes to hours of discharge time. Currently, capacitors and superconducting magnetic energy storage (SMES) are rapid response technologies capable only of providing short-term discharge. Research efforts for both technologies are focused on increasing energy density and decreasing cost, with capacitor efforts being directed in part towards vehicle applications. While SMES research has been active historically, current efforts are modest and there is no clearly defined pathway for SMES to be competitive for applications requiring extended discharge.

Other grid applications require devices with up to about one hour of discharge to provide services such as frequency regulation service (responding to random, rapid variations in demand) and contingency reserves (rapidly responding to a generator or transmission failure). Longer-term storage can also support renewables integration by providing the subhourly ramping requirements which will increase as greater amounts of variable generation sources are added to the grid. Flywheels have been deployed in significant demonstration projects providing frequency regulation. Several battery types have been demonstrated for both frequency regulation and operating reserves, including lithium-ion and various aqueous batteries (such as lead-acid, nickel-cadmium, and nickel-metal hydride). Most aqueous chemistries are considered mature technologies, but additional improvements are possible, even for 100+ year-old lead acid batteries. Research and development efforts on lithium-ion batteries are focused on reducing cost and weight for transportation applications, but these efforts should have spillover benefits to grid applications. In addition, there are certain lithium-ion configurations that are probably unsuitable for transportation applications but potentially suitable for the grid. A major effort by commercial vendors of rapid response technologies such as flywheels and lithium-ion batteries has been gaining access to markets for frequency regulation and full valuation of the response capabilities of the technology.

Energy Management Applications

Grid storage devices for energy management applications can provide continuous discharge for several hours or more. These devices would be potentially useful for shifting energy during periods of low demand (or high renewable supply) to periods of high demand (or low renewable supply). Many of them can also provide the same services as high power/rapid discharge devices.

Pumped hydro storage (PHS) is the dominant technology in this category with about 22 Gigawatts (GW), equivalent to about 22 large power plants, operating in the United States for decades. PHS has high reliability, high efficiency, and long lifetime, but is dependent on the availability of suitable geologic conditions and requires long development times (~10 years including permitting). Based on siting challenges and environmental opposition, PHS suffers from the perception that these issues will prevent large-scale deployment in the future. However, the actual technical potential is large and the number of proposed plants exceeds the current installed capacity, with many of these proposed plants using "closed-cycle" designs that will not interact with existing water bodies and have the potential to reduce both opposition and licensing times. They may also use variable speed equipment improving their ability to provide rapid discharge services.

Compressed air energy storage (CAES) is technically mature, and often considered the lowest-cost option for "bulk" electricity storage, although only one such facility is deployed in the United States. CAES is a hybrid technology which uses natural gas, and typically requires a large underground formation. Major development efforts for CAES currently underway include demonstrating the technology in bedded salt and porous rock. Use of such geologic formations would open up much more of the country to CAES development. Other research and development activities include work on CAES cycles that do not require natural gas fuel.

Hydrogen and other electricity-derived fuels are possible storage options with the advantage of long-term (even seasonal) storage. They currently are among the least efficient (well under 50%) and more expensive storage technologies available and have yet to be deployed beyond small demonstration projects. Fundamental research efforts are required to decrease the cost and increase the durability of electrolyzers and fuel cells. Most of the historic research on hydrogen has been as an alternative fuel for transportation.

Two classes of batteries are currently the primary candidates for electric grid applications—liquid electrolyte flow batteries and high-temperature batteries. High-temperature sodium-sulfur batteries are the most mature and commercially available, with over 270MW deployed worldwide, including installations in the United States. They also have the advantage of relying on low-cost and abundant materials, although manufacturing costs have limited larger-scale use. Sodium sulfur is the only high-temperature battery deployed at large scale, currently manufactured by a single company in Japan. There are several alternative high-temperature chemistries under various stages of research, development, and commercialization. Flow batteries are in the early stages of development and commercialization, with a few U.S. demonstration projects of vanadium and zinc-bromine technologies, with several other technologies under development.

Thermal energy storage (TES) is often overlooked as an electricity storage technology option because it does not store and discharge electricity directly. However, in some applications, thermal storage can be functionally equivalent to electricity storage with efficiencies exceeding 90%, which is higher than most other storage technologies. There are two primary applications of TES for electricity. The first is storing thermal energy from the sun which is later converted into electricity. The currently deployed storage medium is a relatively low-cost molten salt. The primary limitation is that TES is tied to a specific application, in this case concentrating solar power (CSP), which has the challenges of high cost and limited deployment locations, mostly in the desert southwest in the United States. The key research efforts include developing storage materials with higher working temperature, which, when combined with higher temperature CSP plants, will increase efficiency and decrease costs. CSP with thermal energy storage has been deployed in Spain. Construction of a 250 MW CSP/TES facility in the United States is expected to begin in 2012. The second application of TES is cold and hot storage in buildings. Cold storage, used to reduce peak demand from air-conditioning, has been deployed on a relatively large scale. This is a commercially mature technology that provides firm system capacity at very high round-trip efficiency, with the capability of providing multiple grid services. The primary barrier to deployment is capturing the benefits of this distributed technology in the current regulatory and market environment.

Energy Storage for Transportation Applications

As with grid storage, energy storage for transportation applications can be loosely divided into two primary categories: high power/rapid discharge and high energy/extended discharge. High power devices provide short, rapid discharges for vehicle starting and acceleration. While they cannot provide continuous discharge for electrified transport, they can dramatically improve fuel efficiency, as demonstrated by the current generation of hybrid electric vehicles. Currently deployed technologies for these applications include lithium-ion and nickel-based aqueous batteries. Technologies being explored including capacitors, flywheels, and other battery types. Some of these technologies, such as capacitors, may also be used as a fast-responding "buffer" between the electric drive system and the battery or fuel cell in an electric vehicle (EV).

For high energy applications, where stored electricity is actually used to provide a significant fraction of the driving energy, research and development efforts are currently focused on two technologies—hydrogen and batteries. Conceptually, hydrogen is a simple storage technology, produced by splitting water using electricity (among other options), storing hydrogen on board the vehicle, and then converting it to electricity to drive an electric motor via a fuel cell. (Internal combustion engines could also be used, but the low efficiency of that process is less attractive.) The challenges of a hydrogen-based transportation system include the development of an entirely new fueling infrastructure including hydrogen delivery systems and filling stations, with needed safety standards and protocols. The low volumetric energy density of hydrogen makes storage challenging without extremely high-pressure tanks, or advanced chemical storage still in the early research phase. Finally, fuel cells for vehicles remain expensive, with limited lifetimes. There have been demonstration fuel cell vehicle programs by several major auto manufacturers, with announced plans for commercial deployment as soon as 2015. However, substantial research efforts will be needed to reduce costs and improve performance for many of the technologies needed for large-scale hydrogen based transportation. There are other electricity-to-fuel pathways under consideration, but with limited research and development efforts in the United States. They face similar challenges of requiring new fuel infrastructure and currently face much higher costs than fossil fuel alternatives.

The primary alternative to electricity-based fuel production is battery electric storage in plug-in hybrid electric vehicles (PHEVs) and EVs. Most commercially available and proposed EVs and PHEVs (such as the Chevrolet Volt and Nissan Leaf) use lithium-ion batteries. Research and development efforts are focused primarily on reducing cost and increasing energy density as well as safety of lithium-ion technology. Earlier deployed technologies, such as lead-acid used in older EVs and nickel metal hydride used in current HEVs, are not considered likely candidates in future EVs due to fundamental limits of energy density. Concerns have been expressed about the large-scale availability of several metals used in lithium-ion batteries, as well as its concentration in a few geographic regions. In the longer term, lithium-metal and metal-air batteries are in the research and development phase, with the potential of much higher energy density than currently available battery types.

Chapter 2: Background and Scope

In the United States, there are two major motivations for deploying energy storage technologies. The first is to improve the technical and economic performance of the electric power grid ("the grid").[7] This includes enabling more efficient utilization of conventional power plants (e.g., coal- and gas-fired) supplying the grid through load-leveling and providing fast response grid support functions ("ancillary services"), among other services. It also includes enabling greater use of renewable energy sources such as wind and solar generation, which have variable output due to changing weather conditions. Electricity storage is a potential source of grid flexibility to ease integration challenges and decrease integration costs for these renewables.

The second motivation for energy storage is to enable greater use of electrified transportation. The United States is largely self-sufficient for its electricity needs, and has substantial potential to increase production of low-carbon, domestically sourced electricity from renewable and nuclear sources (or from coal using carbon capture). Yet many of these sources cannot directly produce the liquid fuels generally used in conventional vehicles. Electricity storage in batteries or some other technology (including electricity-derived fuels such as hydrogen) could provide a pathway to more electrically powered vehicles, and thereby to reducing U.S. dependence on petroleum.

This report provides information and analysis about the current status and future opportunities for energy storage technologies in electric grid and electric vehicle applications. It attempts to identify technologies which may have a key role in achieving the objectives stated above. It discusses key technical and market barriers, along with research and development (R&D) and policy efforts to reduce those barriers. The report:

- Describes and discusses briefly how existing storage technologies work, their most likely applications, and their advantages for particular applications.

- Describes current limitations of each technology and whether those limitations might be addressed through R&D efforts.

- Describes economic or materials barriers that might impede development or deployment (e.g., requirements for imported precious metals).

- Assesses the costs (fixed and operational), safety, and effectiveness of each technology.

- Assesses the time horizon for market readiness of each technology.

- Provides a technical overview and status of R&D activities for new and emerging storage technologies.

Organization of This Report

This report contains 12 chapters, starting with Chapter 1, "Executive Summary," which provides an overview of the report's main findings. Context and background are provided in Chapter 2, "Background and Scope." Chapter 3, "Overview of Storage Technology Applications and

---

[7] In this report, the electric power grid, or "the grid," refers to the electric power transmission and distribution (T&D) network operated by electric utilities to deliver electricity from generation facilities (including storage) to end users.

Benefits," provides an overview of electricity storage applications and value, including their use to enable renewable electricity; current barriers to deployment; and current initiatives to address technical, economic, and market barriers. Chapters 4-13 discuss each individual storage technology, including a general overview, status in the marketplace (including proposed projects), estimates of current performance, service lifetime, and costs. They also discuss the status of R&D, including key research needs to enable improvements in cost and performance, as well as non-technical barriers including environmental challenges, availability of raw materials, and safety. Each of Chapters 4-13 can be read independently, but Chapters 2 and 3 offer a more complete understanding of some of the more technologically focused discussions that follow.

Chapter 3: Overview of Storage Technology Applications and Benefits

Energy Storage for Electric Power Grids

Electric utilities have long been interested in energy storage technology because of its potential to support the operation of electric power grids. Historically, one of the most important grid storage functions has been "load-leveling," or storing off-peak electricity during periods of low demand and releasing it during periods of high demand, enabling the decreased use of high-cost peaking generation. This function has been extended to include support for renewable electricity generation, given the variable production output of wind and solar plants. More recently, utilities have also been considering how energy storage can provide a partial alternative to the development of the power grid itself by helping utilities optimize the use of grid infrastructure already in place and thereby avoid or defer building new power lines. Other key storage functions include technical services called "ancillary services" needed to provide electric power transmission service to a customer. They include actions taken to effect a power transaction (e.g., scheduling), services needed to maintain the integrity of the power grid, and services needed to correct the effects associated with undertaking a power transaction (e.g., supply-demand balancing).[8] As the electric power grid has evolved into a wholesale marketplace for competitive bulk power purchases while at the same time becoming strained by growth in electricity demand, the potential for energy storage has grown in importance, driving continued interest in storage technology development and deployment.

Current Storage Deployment for the Grid

There are approximately 22 GW of utility-scale electric storage capacity in the United States today, which equates to approximately 2% of the nation's total existing generation capacity.[9] Nearly all of this storage capacity is in the form of pumped hydro storage (PHS), which works by pumping water from a lower reservoir to an upper reservoir, releasing that stored water through a hydroelectric generator when electricity is needed (further discussed in Chapter 9). While there was some development of PHS starting as early as the 1920s, much of the nation's PHS capacity was initiated in the mid- to late 1970s.[10] This development was the result of a combination of factors including dramatic price increases in oil and natural gas used for meeting peak electricity demand, along with concerns about security of supply. These factors culminated in congressional passage of the Powerplant and Industrial Fuel Use Act of 1978 (P.L. 95-620) restricting use of oil and gas in new power plants.[11] During this period utilities expected to bring online many new

---

[8] Federal Energy Regulatory Commission, *Promoting Wholesale Competition Through Open Access Non-discriminatory Transmission Services by Public Utilities; Recovery of Stranded Costs by Public Utilities and Transmitting Utilities*, Order No. 888, April 24, 1996, p. 198.

[9] U.S. Energy Information Administration, *Electric Power Annual*, Table 1.2, April 11, 2011, http://www.eia.gov/cneaf/electricity/epa/epat1p2.html.

[10] American Society of Civil Engineers (ASCE),Task Committee on Pumped Storage of the Hydropower Committee of the Energy Division of the American Society of Civil Engineers, "Compendium of Pumped Storage Plants in the United States," American Society of Civil Engineers, New York., 1993.

[11] U.S. Energy Information Administration, "Repeal of the Powerplant and Industrial Fuel Use Act (1987)," web page, October 18, 2011, http://www.eia.doe.gov/oil_gas/natural_gas/analysis_publications/ngmajorleg/repeal.html.

coal-fired and nuclear power plants to meet relatively steady baseload demand, but were left with limited options to provide generation capacity to meet daily and hourly load variations ("load-following") and peak demand.[12] This limitation led utilities to actively develop PHS as an alternative to fossil-fueled intermediate load and peaking generation.

During the 1970s, there was also significant research and development of other storage technologies including several battery types, capacitors, flywheels, compressed-air, underground pumped hydro, and superconducting magnetic storage.[13] It was expected that deployment of storage of all types would grow significantly during this period.[14] However, most PHS development, along with interest in and deployment of other emerging storage technologies, ended in the 1980s after steep natural gas price reductions, improvements in natural gas turbines, and repeal of the Industrial Fuel Use Act made deployment of flexible natural gas-fired generation more economically attractive.

Other technical, market, and regulatory factors have also served to limit the deployment of electricity storage historically. Many of these factors continue today. They are discussed in more detail later in this chapter and in the individual technology chapters. Briefly, however, a primary historical challenge of storage deployment has been the limited ability of utilities to estimate and capture the full economic value of electricity storage, especially the many dynamic benefits to the grid of fast responding storage technologies.[15] Taken together, these factors have restricted deployment of utility-scale electricity storage in the United States over recent decades. Besides 22 GW of PHS, deployment has been limited to a single 110 MW compressed-air energy storage (CAES) facility, and a variety of smaller projects. Between 1990 and 2010, only 2 MW of new PHS was constructed in the United States compared to over 300 GW of new generating capacity.[16]

Applications of Energy Storage in the Grid

As noted above, energy storage can be used in many valuable applications for electric power grids. A 2010 assessment by Sandia National Laboratories, for example, lists 17 distinct applications and 26 associated benefits of electricity storage.[17] **Table 2** lists some of the most commonly cited applications for electricity storage with a basic description of each. It does not include other possible applications of electricity storage such as "black start" (providing power to restart the grid after a blackout), power quality, voltage and transmission support, substation on-

---

[12] Concern about the availability of oil and other peaking fuels in this period was so great that a 1979 international conference on the subject, which included the U.S. National Academy of Sciences, described energy storage as "a vital element in mankind's quest for survival and progress." J. Silverman,(ed). "Energy Storage: A Vital Element in Mankind's Quest for Survival and Progress," Transactions of the First International Assembly held at Dubrovnik, Yugoslavia, 27 May-1 June, 1979, Pergamon Press, 1980.

[13] U.S. Department of Energy, "DOE Interagency Coordination Meeting on Energy Storage," CONF 7709116, 1977.

[14] D.W. Boyd, O.E. Buckley, and C.E. Clark, "Assessment of Market Potential of Compressed-Air Energy-Storage Systems," *Journal of Energy*, 1983, No. 7, pp. 549-556.

[15] P. Denholm, E. Ela, B. Kirby, and M. Milligan, *The Role of Energy Storage with Renewable Electricity Generation*, NREL/TP-6A2-47187, National Renewable Energy Laboratory, Golden, CO, 2010.

[16] U.S. Energy Information Administration, *Annual Energy Review*, October 19, 2011, Table 8.11a, http://www.eia.gov/totalenergy/data/annual/showtext.cfm?t=ptb0811a.

[17] J. Eyer and G. Corey, *Energy Storage for the Electricity Grid: Benefits and Market Potential Assessment Guide: A Study for the DOE Energy Storage Systems Program*, SAND2010-0815, Sandia National Laboratories, February 2010.

---

site power, and supplemental reserves. Nor does it include the role of storage in supporting variable generation, like wind and solar generation, which is discussed in the next section.

Table 2. Major Power Grid Applications of Electricity Storage

Application	Description	Timescale of Operation
Load Leveling/ Arbitrage/ Time-Shift[a]	Purchasing low-cost off-peak energy and selling it during peak periods with high prices.	Response in minutes to hours. Discharge time of hours.
Firm Capacity	Provide reliable generation capacity to meet peak system demand.	Must be able to discharge continuously for several hours or more.
Operating Reserves		
• Regulation Service	Fast responding increase or decrease in generation (or load) to respond to random, unpredictable variations in demand.	Unit must be able to respond in seconds to minutes. Discharge time is typically minutes.
• Contingency Spinning Reserve[b]	Fast responding increase in generation (or decrease load) to respond to a contingency such as a generator failure.	Unit must begin responding immediately and be fully responsive within 10 minutes. Must be able to hold output for 30 minutes to 2 hours depending on the market.
Ramping/ Load Following	Follow longer-term (hourly) changes in electricity demand.	Response time in minutes to hours. Discharge time may be minutes to hours.
Transmission and Distribution Replacement and Deferral	Reduce loading on electric power grid during peak times. Provides an alternative to expensive and often difficult to site power lines and substations.	Response in minutes to hours. Discharge time of hours.
End-Use Applications		
• Time of Use (TOU) Rates	Functionally the same as arbitrage, just at the customer site.	Same as arbitrage.
• Demand Charge Reduction	Functionally the same as firm capacity, just at the customer site.	Same as firm capacity.
• Backup Power/ Power Quality/ Uninterruptible Power Supply	Functionally similar to contingency reserve, just at the customer site.	Instantaneous response. Discharge time depends on level of reliability needed by customer.

Source: P. Denholm, et al., *The Role of Energy Storage with Renewable Electricity Generation*, NREL/TP-6A2-47187, National Renewable Energy Laboratory, 2010.

a. Arbitrage, strictly defined, is the simultaneous purchase and sale of the same commodity to take advantage of price differences in two different markets. Eyer and Corey (2010) consider the term arbitrage a misnomer as applied to energy storage, however, its use is very common and it is used in this report as well.

b. Contingency reserves may be provided by both spinning and non-spinning units, depending on the market. The requirements for non-spinning reserves are the same except the resource does not need to begin responding immediately, but still requires full response within 10 minutes. These requirements depend upon market and market reliability rules. For an example see PJM, *PJM Manual 11: Scheduling Operations*, Revision 43, September 24, 2009, http://www.pjm.com/markets-and-operations/ancillary-services/~/media/documents/manuals/m11.ashx.

The applications in **Table 2** can be divided any number of ways into a number of different groups. However, for purposes of assessment in this report, it is helpful to classify these applications into two general categories based on the amount of time the storage device is required to provide service because discharge time is a fundamental characteristic distinguishing most energy storage technologies. The first category is storage for *high power* or *rapid discharge* applications where the device must be able to discharge for periods of up to about one hour. The second category is *energy management* applications where the device must be able to discharge for several hours or more. Some applications may overlap both categories, depending upon the type of installation in question, but these categories offer a valid basis for comparison of storage technologies performing, at least to the first degree, similar functions.

Valuation of Storage for the Grid

Each application of electricity storage for the power grid offers distinct benefits. One of the challenges facing electricity storage technologies is appropriate valuation of these benefits, especially in providing multiple services in combination. For example, some storage technologies can provide load-leveling (and associated benefits such as lower cycling-induced maintenance), regulation service, contingency reserves, and firm capacity. Historically, it has been difficult to quantify these various value streams without sophisticated modeling and simulation methods. The emergence of wholesale electricity markets now provides more transparent data for both utilities and independent power producers to consider the opportunities for electricity storage.[18] Depending on the market, these data allow evaluation of both the economic yield and optimum location of electricity storage devices for arbitrage, capacity, operating reserves, and other ancillary services.[19]

---

[18] As of 2009, wholesale energy markets exist in parts of more than 30 states and cover about two-thirds of the U.S. population. Independent System Operators and Regional Transmission Organizations Council, *2009 State of the Markets Report*, prepared by the ISO/RTO Council, 2009, http://www.isorto.org/atf/cf/%7B5B4E85C6-7EAC-40A0-8DC3-003829518EBD%7D/2009%20IRC%20State%20of%20Markets%20Report.pdf.

[19] R. Sioshansi, P. Denholm, T. Jenkin, and J. Weiss, "Estimating the Value of Electricity Storage in PJM: Arbitrage and Some Welfare Effects," *Energy Economics*, No. 31, 2009, pp. 269-277.

Figure 1. Estimated Life-Cycle Value of Several Electricity Grid Storage Applications

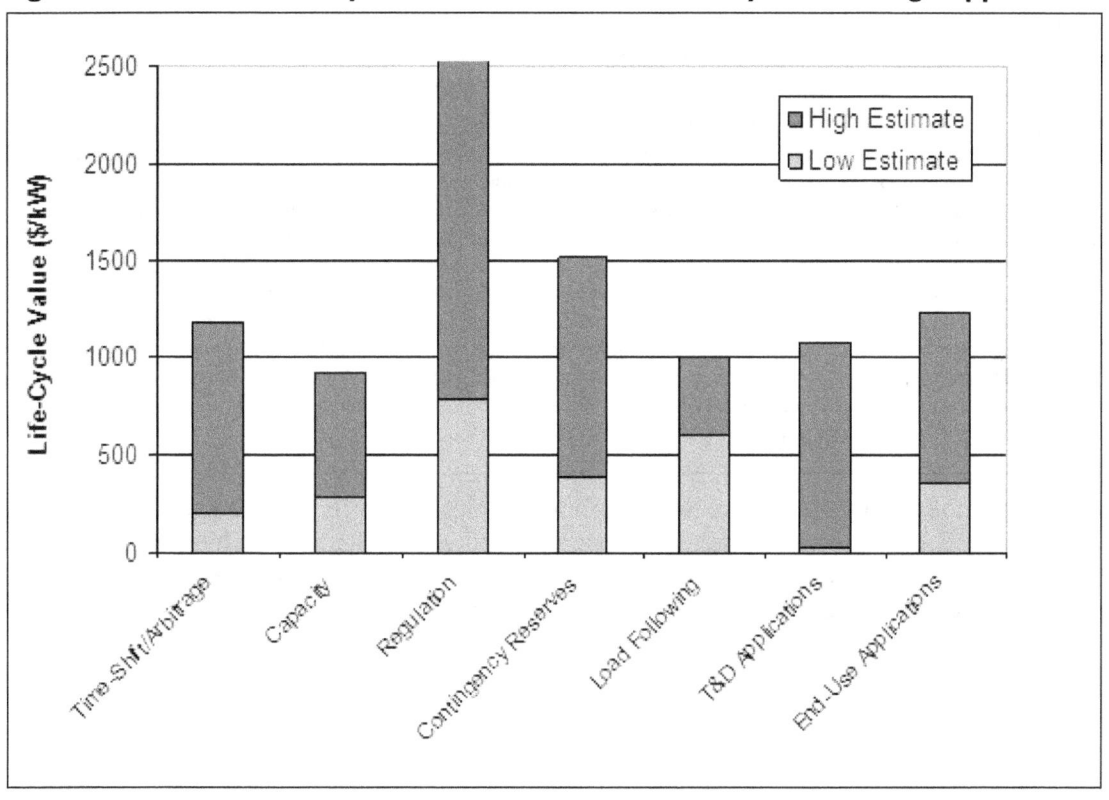

Source: Compiled from Eyer and Corey, 2010 and Denholm *et al.*, 2010.[20]

While there is significant variation and uncertainty in costs, most electricity storage assessments indicate that few commercially available bulk electricity storage technologies are deployable for less than $1,000/kW. For comparison, **Figure 1** summarizes the life-cycle value of several storage applications estimated in several previous studies in a number of locations. (The life-cycle value can serve as a proxy for the capital cost needed for storage to be economically viable for each of these applications.) As **Figure 2** shows, $1,000/kW falls within the range of estimated life-cycle value for all but one of the applications shown, indicating that the storage value could exceed storage costs in specific applications. For example, energy arbitrage revenues, independent of other storage benefits, would require a capital cost of less than $1,000/kW over most of the locations studied. The value of electricity storage increases, however, when taking advantage of other individual sources of revenue or even combined services. A device with sufficient energy capacity for energy arbitrage would likely be able to provide system capacity as well. The combination of these two services could, therefore, likely support a device costing somewhat more than $1,000/kW.

Figure 1 shows that regulation service and contingency reserves, which are shorter-duration applications requiring less energy capacity, have potentially higher value than energy arbitrage. The challenge for these applications is that a device providing contingency reserves must be able to respond rapidly, which is technically harder to do. Frequency regulation is particularly demanding, requiring continuous changes in output, frequent cycling, and fast response. It is also the highest-value opportunity for an electricity storage device, however, and has been the focus of

---

[20] Both studies provide detailed explanation of sources and methods. Regulation value may exceed $4000/kW.

many potential electricity storage applications, especially given its fairly small energy requirements.[21]

Defining the Cost of Electricity Storage

When discussing the costs of storage technologies, a critical issue is that storage devices in electric applications have both a *power* component (kW of discharge capacity) and an *energy* component (kWh of discharge capacity, which may also be expressed as hours of discharge at rated output). The total cost of a storage application must account for the ratings of both components, and may be expressed differently depending on the application or audience. Utilities, for example, universally define the cost of power plants only in terms of rated power ($/kW), so they would expect to see costs in these terms, with the hours of storage (kWh capacity) expressed separately. A grid storage plant might, therefore, be expressed as costing $2000/kW for a device with eight hours of discharge capacity. On the other hand, the battery community typically expresses costs in terms of rated energy ($/kWh), and may or may not include the power component in the cost. So the cost of a battery might be stated as $500/kWh with the power capacity of the battery established separately. When evaluating the economics of storage technologies, care must, therefore, be taken to ensure that the costs for meeting both kW and kWh specifications are included and that both components are "sized" properly for any specific application.

It is difficult to estimate the total market size for electricity storage in the U.S. grid. The most comprehensive assessment of market size identifies hundreds of gigawatts of total applications.[22] However, some of these applications overlap. For example, end use time-of-use (TOU) rate management effectively duplicates load-leveling on the wholesale side. Furthermore, several of the highest-value services, such as regulation service, have the smallest market opportunities.[23] Even with these considerations, the potential market for electricity storage is large, and that market is expected to grow in value and size with the increasing deployment of renewable energy sources.

---

[21] Frequency regulation theoretically is a net zero energy service over relatively short time scales, meaning the energy capacity of the device can be much smaller than that of devices providing operating reserves and energy arbitrage. Several power markets in the United States have changed or have proposed to change their treatment of regulation to accommodate energy-limited storage technologies. Furthermore, it has been suggested that fast-responding storage devices could receive a greater value per unit of capacity actually bid, because they could actually reduce the amount of reserves needed. For example, "faster responsive resources can help to reduce California ISO's regulation procurement by up to 40% (on average)" and "California ISO may consider creating better market opportunities and incentives for fast responsive resources." Y.V. Makarov, J. Ma, S. Lu, and T.B. Nguyen, "Assessing the Value of Regulation Resources Based on Their Time Response Characteristics," Pacific Northwest National Laboratory, PNNL – 17632, June 2008.

[22] Eyer and Corey, 2010.

[23] Requirements for frequency regulation resources are typically set for each Independent Service Operator (ISO) or utility. Regulation requirements frequently change by the month, day, and hour. However, regulation requirements are about 1% of peak capacity, based on NYISO and ISO-NE regulation requirements for 2009. 1% of peak capacity in the entire United States corresponds to about 10 GW.

Energy Storage and Renewable Energy

Renewable energy sources, such as wind and solar generation, create additional opportunities for energy storage deployment due to the variability and uncertainty of the electricity they produce. As variable renewable generation (VG)[24] from these sources is added to the grid, it can have a number of operational impacts on the grid, many of which can be mitigated with electricity storage (or other enabling technologies):

- **Frequency Regulation Requirements**—VG adds to the short-term (seconds to minutes) variability in electric power frequency, which must be maintained very close to the 60 cycles per second (hertz) for proper and reliable grid operation.[25]

- **Load Following Requirements**—VG adds to the hourly requirements for generation supply (ramping) on the grid, increasing the cycling and associated maintenance of conventional generators.

- **Uncertainty in Net Load**—Wind availability is less predictable than either the variation in electric load or the availability of conventional generators. This uncertainty can increase the cost of power system operation because it can result in too many or too few generators being available to respond to variation in "net load," which is the electric load remaining on the grid after wind power supplies are added.[26]

- **Ramping Range and Curtailment**—VG increases the difference between the daily minimum and maximum electricity demand (including an effective reduction in minimum load) which can force conventional generators to reduce output. In some cases this difference may force generation units that ought to be running continuously to cycle off during periods of high wind output, or it can force wind generators to curtail output, "wasting" renewable generation potential.

- **Transmission Requirements**—Some renewable resources, like wind and concentrating solar power, are remotely located, requiring new transmission to supply the grid. New transmission is difficult to construct for economic and

---

[24] The variable generation (VG) nomenclature is used by the North American Electric Reliability Corporation (NERC). See NERC, *Accommodating High Levels of Variable Generation*, April 2009, http://www.nerc.com/docs/pc/ivgtf/ IVGTF_Outline_Report_040708.pdf.

[25] The amount of additional regulation reserves required as a function of VG penetration has yet to be established definitively, especially since the impact on minute-to-minute regulation requirements is mitigated by aggregating large amounts of wind power with variability largely uncorrelated in the regulation time frame. However, a recent analysis by the California Independent System Operator (CAISO) of a 33% renewable portfolio standard suggested that the use of VG could increase regulation requirements by a factor of two to four. See CAISO, *Integration of Renewable Resources: Transmission and Operating Issues and Recommendations for Integrating Renewable Resources on the California ISO Controlled Grid*, November 2007.

[26] This is actually the combination of the uncertainty in load and wind. As VG penetration increases, it begins to dominate the net load uncertainty. This can also result in a shortage of available generation capacity. An example is the ERCOT event of Feb. 26, 2008, where a combination of factors—including greater than predicted electricity demand, forced outage of a conventional generation unit, wind forecast not being given to system operators, and lower than expected wind production—resulted in too little generation capacity online to meet load. As a result, the ERCOT system needed to deploy high-cost quick-start generation units and pay customers to curtail load. This issue has important implications for the use of storage to mitigate uncertainty. Energy storage, like any other generation, must be scheduled; a storage device used for load leveling may not be able to simultaneously provide hedging against under-forecasted wind, because it may already be discharging. See E. Ela, and B. Kirby, *ERCOT Event on February 26, 2008: Lessons Learned*, NREL/TP-500-43373. National Renewable Energy Laboratory, July 2008.

---

policy reasons, however, and use of dedicated long-distance transmission for wind or solar will be limited by the relatively low capacity factor of the resource. Storage could increase line-loading and help reduce wind generation curtailment due to transmission constraints.[27]

Figure 2 illustrates several of the above impacts on net load and corresponding operation of the grid. In this figure, wind generation is subtracted from the load, showing the "residual" or net load that the utility would need to meet with conventional sources. As the figure shows, the change in generation the grid would need to provide for load-following purposes (ramp range) can be much higher than overall load due to the variable contributions of wind power.

Figure 2. Impact on Net Load from Using Wind Generation

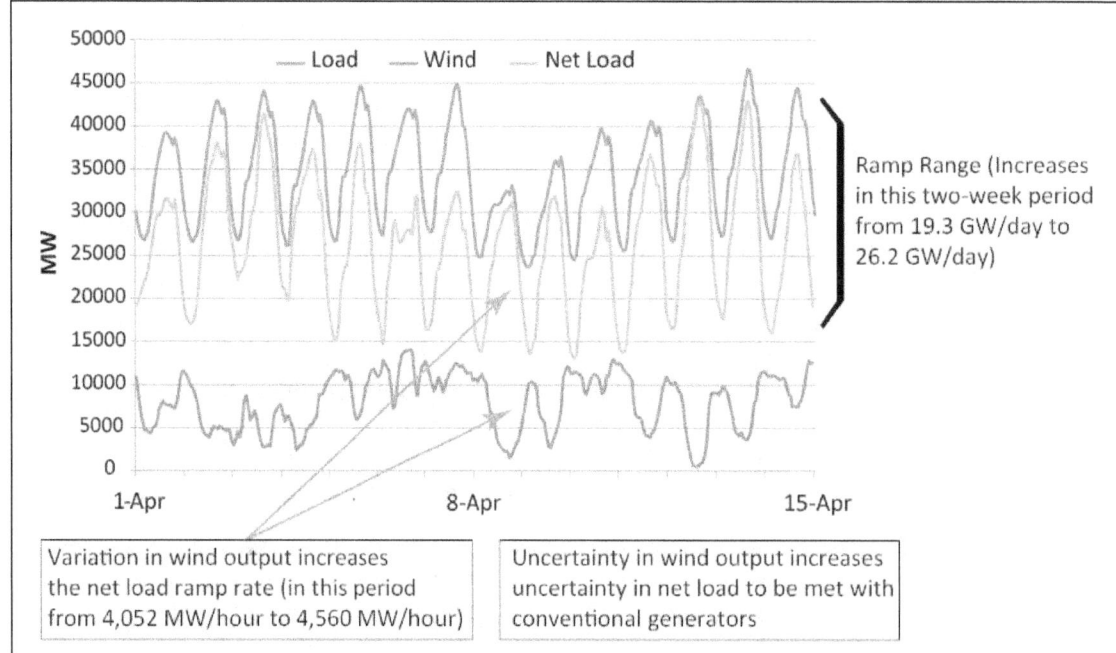

Source: P. Denholm, E. Ela, B. Kirby, and M. Milligan, *The Role of Energy Storage with Renewable Electricity Generation*, NREL/TP-6A2-47187, National Renewable Energy Laboratory, 2010.

Note: This figure uses load data from the Electric Reliability Council of Texas (ERCOT) in 2005 along with 15 GW of spatially diverse simulated wind data from the same year.

Notwithstanding the potential contribution of storage technologies to grid operation, there is considerable debate over the "need" for electricity storage with moderate penetration of renewables. Many of the grid impacts listed above have been evaluated in various wind integration studies attempting to evaluate the operational feasibility and associated costs of wind integration. To date, most studies have found a relatively low cost of accommodating wind

---

[27] Co-locating wind and storage has long been proposed to reduce the amount of transmission needed for new development, coming at the tradeoff of less efficient use of energy storage. See P. Denholm, and R. Sioshansi, "The Value of Compressed Air Energy Storage with Wind in Transmission-Constrained Electric Power Systems," *Energy Policy*, Vol. 37, pp. 3149-3158. This has been proposed to relieve congestion in the Texas grid, for example, where the state's best wind resources are located largely in the sparsely populated western part of the state, and transmission capacity is limited. See N. Desai, et al., *Study of Electric Transmission in Conjunction with Energy Storage Technology*, Lower Colorado River Authority, Texas State Energy Conservation Office, August 21, 2003.

---

variability on the grid—typically less than $5/MWh (0.5 cents/kWh), adding less than 10% to the cost of wind energy—when wind is providing up to 20% of a particular grid system's demand.[28] This low cost, mostly resulting from the large amount of flexible generation already available to meet the variability in demand, has been used to argue that deployment of storage is not justified based on variability impacts.[29] However, many potentially significant costs, such as the operations and maintenance costs of increased generator cycling, have yet to be quantified and are not included in these studies.[30] Furthermore, the studies do not consider the economic and societal challenges associated with transmission expansion, or the option of storage as a method to supplement new transmission.[31] Consideration of these factors would almost certainly increase the value of storage. A strong argument also can be made that VG will increase the value proposition for storage, adding to the values that already exist in today's grid. In general, renewables are likely to increase the potential market size for electricity storage (for example, by increasing the amount of certain types of generation reserves needed).

The operational value of combining a dedicated electricity storage device with a specific wind or solar generation plant is also the subject of debate. Many storage applications dedicated to individual renewable generators, such as renewables "firming," which seeks to reduce variability in renewable power output, are actually specific examples of the more general applications in **Table 2**. For example, shifting wind power supply from periods of low demand to periods of high demand is fundamentally the same as energy arbitrage. The economic benefits of this application are greatest when the electricity storage operator can choose from all of the generators in a system, storing electricity from any source instead of storing only wind generation, when demand is lowest.[32]

---

[28] J. DeCesaro, K. Porter, and M. Milligan., "Wind Energy and Power System Operations: A Review of Wind Integration Studies to Date," *The Electricity Journal*, Vol. 22, No. 10, December 2009, pp. 34-43.

[29] Note that these studies have not necessarily focused on storage and generally do not attempt to determine the optimal system (including the amount of storage if any) that provides the lowest cost of energy.

[30] Wind integration studies typically use proprietary software and data sets, and do not always state which costs are included or excluded. However, the Western Wind and Solar Integration Study (the highest penetration U.S. integration study as of 2009) states: "'Wear and tear' costs due to increased or harder cycling of units were not taken into account because these have not been adequately quantified." See D. Lew et al., "How do Wind and Solar Power Affect Grid Operations: The Western Wind and Solar Integration Study," NREL/CP-550-46517, National Renewable Energy Laboratory, September 2009.

[31] The more recent U.S. studies of very high penetration (the Western Wind and Solar Integration Study and the Eastern Wind Integration Study) require power and energy exchanges over larger areas than typically occur in the existing system. See M. Milligan, et al., *Large-Scale Wind Integration Studies in the United States: Preliminary Results*, NREL/CP-550-46527, September 2009.

[32] There are some exceptions when there are benefits of operationally combining VG and energy storage, typically through co-location and sharing of certain high-cost components. The best example is integrating thermal storage into a concentrating solar power (CSP) plant; another example is sharing power electronics in a distributed PV/battery system. There are several other applications where co-location of VG and storage may make sense. Wind plants placed in areas of weak transmission can potentially introduce power quality and stability issues, and storage can be a mitigating technology; however, improved power electronics in modern wind turbines may be a lower-cost alternative. Finally, combining wind and energy storage has been proposed as an alternative (or supplement) to developing new transmission capacity. Despite these potential applications, the majority of storage deployed in the grid will likely be a shared resource, which will benefit the entire system and not just a single generator or load. J.C. Smith et al., "Utility Wind Integration and Operating Impact State of the Art," *IEEE Transactions On Power Systems*, Vol. 22, No. 3, August 2007. Just as loads are balanced in aggregate, the net load in the future grid—after all VG sources are included—will be balanced by a mix of conventional generation, plus flexibility options that may include energy storage.

Grid Storage with High Renewables Penetration

Perhaps the strongest argument for energy storage in the grid occurs at relatively high penetration of VG. The oft-cited limits of VG penetration in the range of 10%-20% and the associated "need" for electricity storage appear to be moving targets as new grid integration techniques develop. Recent studies have found that 30% penetration (on an energy basis) of renewables on the grid appears to be feasible without an inherent need for storage to maintain system reliability.[33] However, there appear to be some economic limits of VG at high penetration based on the limited coincidence of VG supply and electricity demand patterns.[34] At sufficiently high penetration of wind or solar generation, VG supply can exceed demand for electricity, which results in curtailed generation and decreased economic viability of VG. This problem is exacerbated by the cycling or operational limits on conventional generators, many of which must remain on-line to provide operating reserves or be available when wind and solar generation is insufficient to meet demand.

Such VG limits can be observed in the Western Wind and Solar Integration Study (WWSIS) with 30% wind penetration, at which wind power supply almost completely removes conventional generation during high wind periods.[35] **Figure 3** shows the net load with wind in the study area, along with the modeled operation (dispatch) of generation plants, which requires significant ramping of coal generators. In one evening the net load (electricity demand minus wind supply) drops to about 6 GW, meaning that wind is providing about 32 GW, even after much of the wind generation is exported to surrounding areas. Doubling the amount of wind generation capacity would produce a large amount of wind generation curtailment on this day, since the remaining 6 GW of load cannot absorb an additional 32 GW of wind generation, and the conventional generation units are probably near or at their ability to ramp down or cycle off. Similarly, an analysis of the Irish grid found limited wind generation curtailment at a wind penetration of 40% on an energy basis.[36] However, beyond this point, wind curtailment rates sharply increase and the study found economic benefits of storage at the point where about 50% of the system's energy is provided by wind.

At higher penetration of renewable generation, the ability of conventional generators to reduce output becomes an increasing concern. VG begins to displace units that are traditionally not cycled, and the ability of conventional thermal generators to reduce output may become constrained.[37] Utilities in the United States have expressed concern about their systems "bottoming out" due to the minimum generation requirements during overnight hours, and being unable to accommodate more VG during these periods. Cycling operations, including startup/shutdown, on-load cycling, and high frequency MW changes, can damage generation equipment. However, the costs of such cycling can be very difficult to quantify.[38] Minimum load points would be less of a constraint if conventional power plants could be quickly shut down and

---

[33] M. Milligan et al., *Large-Scale Wind Integration Studies in the United States: Preliminary Results*, NREL/CP-550-46527, National Renewable Energy Laboratory, September 2009.

[34] P. Denholm and M. Hand, "Grid Flexibility and Storage Required to Achieve Very High Penetration of Variable Renewable Electricity," *Energy Policy*, No. 39, 2011, pp. 1817-1830.

[35] GE Energy, *Western Wind and Solar Integration Study*, prepared for National Renewable Energy Laboratory, NREL Report No. SR-550-47434, May 2010.

[36] A. Tuohy and M. O'Malley, "Impact of Pumped Storage on Power Systems with Increasing Wind Penetration," *Energy Policy*, Vol. 39, No. 4, 2011, pp.1965-1974.

[37] For example, see M. Milligan, et al., *The Impact of Electric Industry Structure on High Wind Penetration Potential*, NREL/TP-550-46273, National Renewable Energy Laboratory, July 2009.

[38] S.A Lefton, and P. Besuner, "The Cost of Cycling Coal Fired Power Plants," *Coal Power*, Winter 2006.

started up at low cost. However, with the exception of certain peaking power plants such as aeroderivative turbines and fast-starting reciprocating engines, most conventional plants have minimum up-and-down times, and require several hours to restart—at considerable cost.

Figure 3. Generation Dispatch in the WWSIS Study at 30% Wind Penetration

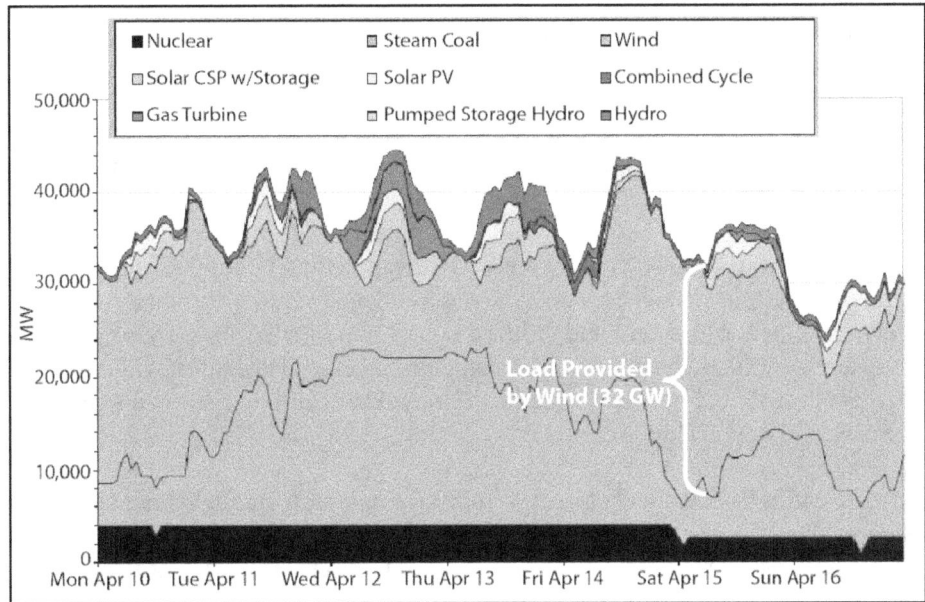

Source: GE Energy (for NREL), 2010.

In some markets, electricity prices have dropped below the actual variable (fuel) cost of producing electricity on a number of occasions. This indicates that power plant operators are willing to sell energy at a loss to avoid further reducing output. At this point an increasing fraction of wind generation will simply be unusable by the system and electricity storage becomes an increasingly attractive method of shifting otherwise curtailed wind generation to times of lower wind generation (and/or higher loads). Overall, the increase in energy storage value or market size associated as a function of increasing VG penetration is not well quantified. In addition, financial mechanisms for energy storage installations to recover their costs as VG-enabling technologies are incomplete.

Ongoing Barriers to Storage Deployment for the Grid

Historically, the primary barriers to energy storage deployment for the grid have been establishing a positive benefit/cost ratio for storage and actually capturing the economic value that storage provides. While the emergence of restructured wholesale electricity markets has provided new storage opportunities, electricity storage still faces significant non-technical barriers to widespread market acceptance and adoption, further discussed below.

Unquantified and Uncaptured Benefits

Before wholesale electricity markets began to be restructured in the 1990s, the value of ancillary services and other grid support functions was largely hidden in electric utilities' cost of service.

For example, the value of providing operating generation reserves, which affect the ability of a power plant to respond to the electric grid's dynamic operating needs, was rarely calculated.[39] The Federal Energy Regulatory Commission (FERC), which regulates wholesale transmission grid tariffs, and grid operators are increasingly recognizing the value of these services, and the advantages of electricity storage in providing them, especially because they generally require fast response and limited energy delivery for which storage devices are well-suited. However, much of the nation remains in a traditional regulated utility framework, where the benefits of storage in providing grid support services remain undervalued. Furthermore, wholesale electricity markets do not capture all the costs of generation plant operation, especially those related to cycling and ramping. Quantifying the full value of energy storage remains challenging due in part to the limited ability of utility models to simulate realistic power plant and storage system operation over multiple time scales.

Wholesale electricity markets also do not capture all the potential benefits of storage to the electric distribution system (which connects the high voltage grid to electricity end users), including deferral of new equipment and reduced power line losses.[40] Deploying storage in the distribution system will likely be particularly challenging since distribution will almost certainly remain a regulated monopoly utility service, with limited exposure to market conditions that provide incentives for new technologies.

Finally, there are currently few mechanisms in place for potential energy storage operators to capture economic benefits associated with enabling renewable energy sources. Some value may be captured indirectly. For example, if VG increases regulation service requirements, then it also increases market opportunities for storage. However, there are few comprehensive mechanisms to capture any potential synergies between VG and storage.

Regulatory and Market Uncertainty and Risk

Utilities tend to be risk averse. To meet electricity supply requirements, they tend to rely on mature generation technologies with which they have long-term experience rather than new technologies. Conventional generation options, including flexible natural gas-fired turbines, continue to be the primary option for load following, peak power generation, and ancillary services. Market uncertainty, combined with a lack of incentives for risk taking in regulated utilities, discourages the deployment of technologies that are new or have long lead times. Long development times and risk are a particular challenge for the two leading options for bulk energy storage—compressed air and pumped hydro. PHS, in particular, faces unique environmental and other siting challenges (including new transmission requirements), and also faces long permitting and construction times. The regulatory treatment of energy storage for the grid is often unclear, and has complicated the financing of large storage projects.[41] These issues are discussed in more detail in the technology sections.

---

[39] S.J. Jabbour, and W.M. Wells, "Optimal Dispatching of Storage Plants with Dynamics," *Proceedings of the Second International Conference on Compressed Air Energy Storage*, EPRI TR-101770, Electric Power Research Institute, December 1992.

[40] A. Nourai, V. I. Kogan, and C.M. Schafer, "Load Leveling Reduces T&D Line Losses," *IEEE Transactions on Power Delivery*, Vol. 23, No. 4, October 2008, pp. 2168-2173.

[41] A recent example is the Lake Elsinore Advanced Pumped Storage Project, which applied to be considered a transmission facility for purposes of utility rate recovery. FERC denied this request, forcing it to recover costs through an alternative mechanism, such as the more risky (at least for the developer) generation market. Federal Energy (continued...)

Lack of Incentives for Customer-Sited Storage

As with deployment by utilities or independent power producers, customer-sited storage faces challenges of valuation and capturing that value. The benefits of customer-sited storage can exceed that of centrally deployed storage. In addition to providing load-leveling and ancillary services, customer-sited storage can provide additional advantages of reduced distribution losses and increased grid capacity. Some customers, particularly large commercial and industrial consumers, can capture some of the benefits of load-leveling and peak capacity via time-of-use or demand-based electricity rates. But many storage benefits, particularly the value of ancillary services, cannot be captured through their rates. This makes electricity storage uncompetitive for many electricity end users.

In summary, energy storage for the grid faces significant barriers to being evaluated on the same economic terms as conventional grid options perceived to be less risky for utilities both in restructured markets and in traditional integrated utilities. Storage also faces increasing competition from a variety of technical and market options for providing grid flexibility. New market mechanisms are being deployed to share generation, reserves, and net loads—all of which can increase overall power system flexibility.[42] Demand response also may compete against new storage options as a significant source of operating reserves. In many locations in the United States, demand is increasingly used as a source of grid services. In Texas demand response typically provides half of the contingency reserve requirements. Other regions also use (or are evaluating) load to provide regulation. Greater participation of load providing reserves and load shifting will require regulatory and policy changes in addition to new technologies.

Current Grid Storage Policies

Recognition of the potential value of energy storage for grid applications has led to efforts by federal and state agencies to engage in storage R&D efforts ranging from analysis of benefits to providing direct incentives.

Analysis of Storage Benefits

Federal and state agencies have supported a number of studies to evaluate the potential role and value of energy storage. These studies have demonstrated the potential benefits of traditional storage applications discussed above.[43] Other analyses have identified the unique benefits of fast response electricity storage technologies (e.g., flywheels) in providing frequency regulation more efficiently and with fewer emissions than conventional generation.[44] Due in part to such analysis,

---

(...continued)

Regulatory Commission, "FERC Encourages Transmission Grid Investment," Docket No. ER06-278-000, March 20, 2008.

[42] Greater aggregation of loads and reserves has historically been one of the least-cost methods of dealing with demand variability, especially because it often requires operational changes and relatively little new physical infrastructure. This includes introducing sub-hourly markets that allow systems faster response to variability. See M. Milligan, et al., "The Impact of Electric Industry Structure on High Wind Penetration Potential," NREL/TP-550-46273, National Renewable Energy Laboratory, July 2009.

[43] Many of these studies have been performed by the Department of Energy's Energy Storage Systems Research Program, managed by Sandia National Laboratories.

[44] Makarov et al. 2008. As noted earlier the ramp requirement of 1 MW/min could easily be provided by a 1 MW (continued...)

there is a growing consensus, for example, that fast-response resources should be paid a premium for regulation services in wholesale power markets to accurately reflect the value they add to the electric system.[45] As the Chairman of the Federal Energy Regulatory Commission has stated,

> Regarding compensation, some storage technologies appear able to provide a nearly instantaneous response to regulation signals, in a manner that is also more accurate than conventional resources. These two characteristics can reduce the size, and hence overall expense, of the regulation market. Most existing tariffs or markets do not compensate resources for superior speed or accuracy of regulation response, but such payment may be appropriate in the future....[46]

Other studies have demonstrated the potential benefit of electricity storage in supporting the deployment of variable generation[47] and reducing distribution losses.[48] Analytic efforts like these help guide policy and market reforms to appropriately capture the value of grid storage services.

Research, Development, and Demonstration Projects

Federal and local agencies have supported basic research, engineering, technology development, and demonstration programs for many energy storage technologies in grid applications. Until recently, Department of Energy (DOE) R&D efforts in grid electricity storage were relatively modest. From 1992 through 2008 the annual budget for the Energy Storage Systems Program within the DOE's Office of Electricity Deliverability and Energy Reliability was typically less than $10 million per year.[49] (Programs supporting storage primarily for transportation applications are discussed in the next section.) In 2010 the DOE budget was increased to $14 million. The American Recovery and Reinvestment Act of 2009 (ARRA) also greatly increased funding for storage R&D through several programs. Applied research has been supported through the Advanced Research Projects Agency–Energy (ARPA-E) program, with $30.6M awarded for FY2010 and $37.7 awarded for FY2011.[50]

(...continued)

flywheel, but would require about 2 MW of hydroelectric capacity, 3 MW of gas-fired combustion turbine capacity, or 30 MW of gas-fired combined cycle or coal capacity. As a result, using fast responding energy storage to provide regulation can reduce the amount of regulation required, potentially reducing system costs.

[45] Federal Energy Regulatory Commission, *Order Accepting Tariff Revisions*, Docket ER09-836-000, May 15, 2009. http://www.nyiso.com/public/webdocs/documents/regulatory/orders/2009/05/ FERC_Ordr_NYISO_Intgrtd_LESRs_NYISO_DAM_RTM_05_15_09.pdf

[46] Jon Wellinghoff, Chairman, Federal Energy Regulatory Commission, Testimony before the Senate Committee on Energy and Natural Resources Hearing on Energy Storage, Dec. 10, 2009.

[47] KEMA, Inc., *Research Evaluation of Wind Generation, Solar Generation, and Storage Impact on the California Grid*, prepared for the California Energy Commission, Public Interest Energy Research Program, CEC-500-2-1-010, June 2010.

[48] A. Nourai, V.I Kogan, and C.M. Schafer, "Load Leveling Reduces T&D Line Losses," *IEEE Transactions on Power Delivery*, Vol. 23, No. 4, October 2008, pp.2168-2173.

[49] J. Boyes, "FY07 DOE Energy Storage Program Peer Review," Sandia National Laboratories, slide presentation, 2007. http://www.sandia.gov/ess/docs/pr_conferences/2007/boyes_snl.pdf

[50] M. Johnson , "Gridscale Rampable Intermittent Dispatchable Storage (GRIDS) Program," presentation to the DOE Annual Storage R&D Review Meeting, November 2010. http://www.sandia.gov/ess/docs/pr_conferences/2010/ johnson_doe.pdf

Electricity storage demonstrations have been funded directly through the ARRA with a total funding of $185 million.[51] Demonstration programs are particularly important in the electric utility sector, since regulated utilities are typically not rewarded for risk taking, and have few incentives to be the first to deploy new technologies. These funding activities are discussed in more detail in the technology chapters. State agencies have also supported electricity storage demonstrations, often with co-funding from federal agencies. Examples include New York State Energy Research and Development Authority (NYSERDA) support of demonstration programs for flywheels and several battery technologies.[52] The California Energy Commission (CEC) also has supported at least 20 storage research and demonstration projects since 1990.[53]

Market Rules

Securing the ability of energy storage to compete on common terms against traditional generation assets is a critical challenge for grid storage developers. The creation of wholesale markets allows increased participation of electricity storage devices, but the level of participation varies by market. In 2007 FERC issued Order 890 requiring wholesale markets to consider non-generation resources for grid services. The order required that non-generation resources (including energy storage and demand response) be evaluated on a comparable basis to services provided by generation resources in meeting mandatory reliability standards, providing ancillary services, and planning the expansion of the transmission grid.[54]

Since that time Independent System Operators (ISOs) and Regional Transmission Organizations (RTOs), the entities that operate regional power grids, have increased market access, including creating new tariffs for electricity storage.[55][56] In October 2011, FERC issued Order 755 requiring a new compensation method for grid regulation service "to remedy undue discrimination" against faster-ramping resources such as energy storage.[57]

Several large-scale grid storage projects have been proposed or constructed to take advantage of high-value ancillary service markets. Examples of operating projects include a 20 MW flywheel facility in New York and a 12 MW battery facility in Chile.[58] However, market rules are still

---

[51] E. Christy, "Energy Storage Systems Program: 2010 Update Conference" National Energy Technology Laboratory. November 2, 2010. http://www.sandia.gov/ess/docs/pr_conferences/2010/christy_doe.pdf

[52] G. Huff, "NYSERDA/DOE Joint Energy Storage Initiative," Sandia National Laboratories, November 2, 2010. http://www.sandia.gov/ess/docs/pr_conferences/2010/huff_snl.pdf

[53] P. Kulkarni, "California Energy Commission Support for Electricity Energy Storage," California Energy Commission, May 6, 2009. http://bscleantech.org/bscc3/presentations/Technology%20-%20CEC%20-%20Pramod%20Kulkarni.pdf

[54] Federal Energy Regulatory Commission, *Preventing Undue Discrimination and Preference in Transmission Service*, Order No. 890, February 16, 2007.

[55] For example, the New York ISO created a "limited energy storage resource"(LESR) tariff. In its approval of the tariff, FERC stated "We find that the proposed tariff revisions to incorporate LESRs will benefit NYISO's markets by providing them with a new source of regulation service with unique operational characteristics that enable very fast responses to needs for regulation." Federal Energy Regulatory Commission, Order Accepting Tariff Revisions, Docket ER09-836-000, May 15, 2009. http://www.ferc.gov/EventCalendar/Files/20090515142559-ER09-836-000.pdf

[56] Federal Energy Regulatory Commission, *Order Conditionally Accepting Stored Resources Compliance Filing*, Docket No. ER09-1126-001, May 10, 2009.

[57] Federal Energy Regulatory Commission, *Frequency Regulation Compensation in the Organized Wholesale Power Markets*, Order No. 755, October 20, 2011.

[58] Sonal Patel, "Milestones for Flywheel, Lithium Battery Grid-Scale Projects," *Power*, August 1, 2011.

---

evolving in some regions and much of the United States has no access to restructured energy markets to begin with.[59] Uncertainty remains as to how storage assets should be able to capture multiple value streams. Challenges remain in gaining access to distribution and customer-sited storage. One storage company has developed a business model in which customer-sited storage is owned by the utility as a peak generation and load shifting asset.[60]

Incentives for Deployment

There have been a number of financial incentive programs for grid storage technologies offered by the federal government. In addition to the direct funding of demonstration programs, the ARRA amended the DOE's Loan Guarantee Program making certain electricity storage technologies eligible.[61] This program has been applied to a large solar plant in Arizona, discussed in Chapter 12. The ARRA also established a manufacturing tax credit that could potentially be applied to electricity storage manufacturing facilities. Several states have incentives supporting deployment of renewable energy and energy efficiency devices which could be applicable to storage-related equipment including fuel cells and cold thermal storage, but the impact of these programs on actual adoption has been modest. Finally, certain renewable generators are eligible for a 30% federal investment tax credit (ITC), currently scheduled to expire in 2016. This means that thermal energy storage for concentrating solar power is eligible, since it is integrated into a renewable generator. However, stand-alone storage technologies are not covered, since they are typically not integrated into individual renewable generation installations.

A federal direct incentive program was proposed in 2010, which included a 20%-30% ITC for new storage investments depending on size and application.[62] Various other federal energy and climate-change proposals have included language either providing financial incentives for or otherwise encouraging energy storage deployment for the grid, but these proposals have yet to be enacted.

Storage Portfolio Standards

Recently there have been proposals for government-mandated energy storage portfolio standards, similar to renewable portfolio standards (which require utilities to purchase a certain portion of their energy supplies from renewable generators).[63] One example that has been enacted in state law is California's AB 2514, which as originally proposed required certain utilities to install

---

[59] See, for example: "Revised Draft Final Proposal for Participation of Non-Generator Resources in California ISO Ancillary Services Markets." California Independent System Operator, March2010.

[60] Ice Energy, "SCPPA to Undertake Industry's Largest Utility-Scale Distributed Energy Storage Project," press release, January 27, 2010. http://www.ice-energy.com/content10197

[61] The loan guarantee program was created to support the deployment of innovative clean energy technologies pursuant to Section 1703 of Title XVII of the Energy Policy Act of 2005. Title XVII was amended by the American Recovery and Reinvestment Act of 2009 to create Section 1705, a new program for deploying renewable energy and electric power transmission projects.

[62] "Storage Technology for Renewable and Green Energy Act of 2010," S. 3617, 111th Cong., 2nd Sess, 2010.

[63] Brian Nese, "Energy Storage Developers Call for National Storage Portfolio Standard," *Renewable + Law*, Internet blog, July 21, 2009. http://www.lawofrenewableenergy.com/2009/07/articles/power-storage/energy-storage-developers-call-for-national-storage-portfolio-standard/

---

storage devices to meet 2.25% of peak demand. As passed, the bill requires the California Public Utilities Commission to determine targets by March 1, 2012.[64]

Storage for Electric Transportation Applications

Transportation Storage Technologies and Pathways

The primary purposes of electrifying transportation are to reduce dependence on oil, which currently provides most of the nation's transportation fuel, and to reduce vehicle emissions. There are two pathways to store electricity for use in electric vehicle (EV) fleets (**Figure 4**).[65] The first is switching from oil-derived fuels to one of several electricity-derived fuels, either gaseous or liquid, with hydrogen receiving the most attention in recent years. These alternative fuels can be produced using electricity (for example, by splitting hydrogen from oxygen atoms in water) either centrally or near the point of use. They can then be burned in a vehicle using a modified internal combustion (IC) engine and a conventional drive train. Such fuels can also be burned in an IC engine-electric drive train (hybrid-electric) vehicle configuration (HEV), or in a similar fuel cell electric vehicle (FCEV) configuration. The second pathway for electrified transport is to store electricity on board the vehicle, primarily using batteries, and to use that stored electricity to power an electric motor. The vehicle can be either a "pure" battery electric vehicle (BEV) or a vehicle that uses both stored grid electricity and an IC or fuel cell engine, typically referred to as a plug-in hybrid electric vehicle (PHEV).

Figure 4. Pathways to Vehicle Electrification

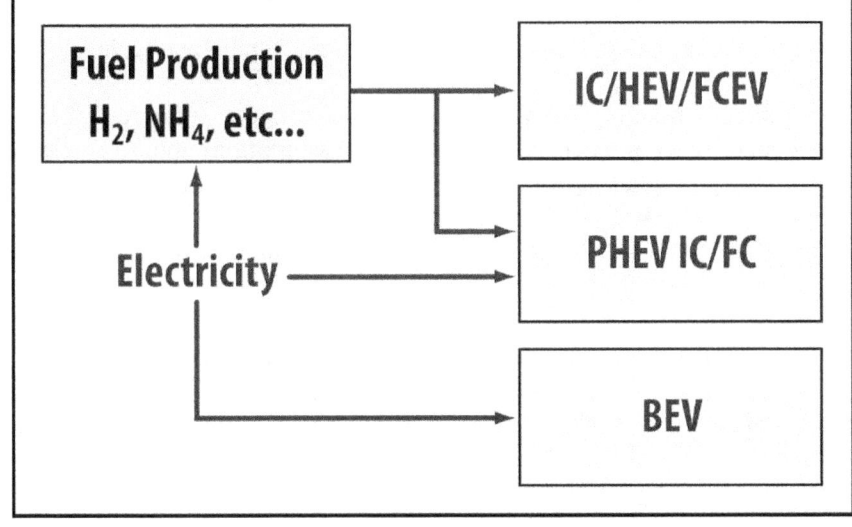

Source: P. Denholm, National Renewable Energy Laboratory.

Key: B = battery, EV = electric vehicle, FC = fuel cell, H = hybrid, IC = internal combustion, P = plug-in.

---

[64] California Legislature, A.B. 2514 (introduced), February 19, 2010. http://leginfo.ca.gov/pub/09-10/bill/asm/ab_2501-2550/ab_2514_bill_20100219_introduced.pdf); and A.B. 2514 (approved), September 29, 2010. http://www.leginfo.ca.gov/pub/09-10/bill/asm/ab_2501-2550/ab_2514_bill_20100929_chaptered.pdf

[65] This report does not consider alternative fuels, such as ethanol, that are not primarily derived from electricity.

Impacts and Benefits of Vehicle Electrification

The most obvious benefit of vehicle electrification is reduced dependence on petroleum-derived fuels. The amount of displaced petroleum depends on the degree of electrification of individual vehicles and of the fleet as a whole. An FCEV running on hydrogen or a pure electric vehicle uses no gasoline, while a PHEV could reduce a large fraction of gasoline use, depending on battery size and driving patterns.

Shifting from gasoline to electricity may have a number of impacts on the electric power grid. One possible outcome is a need for new generation capacity for battery charging. However, the availability of off-peak generation is estimated to be sufficient for a large number of vehicles assuming some level of "smart" charging. A 2007 study estimated that spare generating capacity could have electrified 73% of light-duty gasoline vehicles in 2002.[66] This level of vehicle electrification would displace petroleum equivalent to more than 50% of the nation's oil imports. Another study used a somewhat more conservative methodology to estimate that the bulk power system in 2002 could have supported electrification of approximately 37% of vehicle miles traveled (VMTs).[67] The Electric Power Research Institute (EPRI) analyzed a scenario with 20% of VMTs in the United States powered by electricity in 2030; the modeled electric generating capacity was just 1.7% higher in this scenario compared to the base case scenario that assumed no EVs or PHEVs.[68] Other, regional studies similar have concluded that, essentially, if vehicles charge off-peak, a large number of vehicles can be accommodated, but if on-peak charging is allowed, there could be increased generation requirements during peak periods of electricity demand.[69] On the electric distribution side, the impacts of vehicle electrification are more complex. In some locations, a concentration of vehicle charging could exceed the capacity of distribution systems, and increased loads could shorten the lifetimes of distribution transformers. Distribution system impacts and the need for upgrades, as well as the ability to reduce the impacts via smart charging schemes, will need to be further evaluated, typically on a local level.[70]

EVs and PHEVs generally produce lower greenhouse gas emissions per mile than conventional vehicles. The amount of reduction depends on numerous assumptions about vehicle performance and the mix of electricity supplies used for charging. One estimate is that a PHEV powered by an average proportion of coal-generated electricity produces carbon emissions per mile similar to those of an HEV.[71] If the PHEV is charged using the current grid average emissions, carbon

---

[66] M. Kintner-Meyer et al., *Impacts Assessment of Plug-In Hybrid Vehicles on Electric Utilities and Regional U.S. Power Grids. Part 1: Technical Analysis*, Pacific Northwest National Laboratory, 2007.

[67] The biggest difference between the methods is estimating which generating capacity is available and economical to use to charge vehicles during the peak months. C.H. Stephan and J. Sullivan, "Environmental and energy implications of plug-in hybrid-electric vehicles," *Environmental Science & Technology*, Vol. 42 No. 4, 2008, pp. 1185-1190.

[68] M. Duvall and E. Knipping, *Environmental Assessment of Plug-In Hybrid Electric Vehicles. Volume 1: Nationwide Greenhouse Gas Emissions,* Electric Power Research Institute, 2007.

[69] Examples include: P. Denholm and W. Short, *An Evaluation of Utility System Impacts and Benefits of Optimally Dispatched Plug-In Hybrid Electric Vehicles*, NREL/TP-620-40293, National Renewable Energy Laboratory, 2006; and K. Parks, P. Denholm, and T. Markel., *Costs and Emissions Associated with Plug-In Hybrid Electric Vehicle Charging in the Xcel Energy Colorado Service Territory*, NREL/TP-640-41410, National Renewable Energy Laboratory, 2007.

[70] C. Farmer et al., "Modeling the Impact of Increasing PHEV Loads on the Distribution Infrastructure," 43rd Hawaii International Conference on System Sciences (HICSS), January 5-8, 2010. http://www.cems.uvm.edu/~phines/ publications/2010/farmer_2010_phev_distribution.pdf.

[71] In the worst case scenario for a PHEV, net CO_2 emissions are about the same as those of a conventional vehicle. Any electricity supply mix less than 100% coal-generated will be cleaner. See C.H. Stephan, and J. Sullivan, (continued...)

---

emissions per mile are reduced by almost 60% compared to a conventional vehicle.[72] Air pollutant emission changes due to EV and PHEV penetration are complicated because they depend on the type of generators used for electricity production, the pollution control equipment, and policies that limit emissions. A California study projects that PHEVs would reduce nitrogen oxide (NO_x) and volatile organic compound (VOC) emissions per mile by 87% and 97%, respectively, due to limits on the emissions of these pollutants.[73] Estimates from other regions project changes over a large range, including some locations where net emissions could increase if current power plants do not install new pollution control devices.[74] For example, one study found that life cycle NO_x emissions changes could range from -70% (assuming charging with renewable generation) to +38% (charging with uncontrolled coal-fired plants) assuming no pollution control policies.[75] The actual impact on air quality is even more complex because PHEVs displace emissions from urban areas to rural areas where power plants are typically located and fewer people live. Estimates using air quality models generally indicate improved air quality in urban areas as a result of vehicle electrification.[76]

When parked, vehicles could potentially provide various grid services. Charging of EVs can potentially be controlled and can provide a source of dispatchable demand and demand response. Controlled charging can be timed to periods of greatest VG output, while charging rates can be controlled to provide contingency reserves or frequency regulation reserves. Vehicle-to-grid (V2G) (where EVs can partially discharge stored energy to the grid) may provide additional value by acting as a distributed source of energy storage. Most proposals for V2G focus on short-term response services such as frequency regulation and contingency. Their ability to provide energy services is more limited by both the storage capacity of the battery and the high cost of battery cycling. This could restrict their ability to provide time shifting (energy arbitrage) beyond their ability to perform controlled charging.[77] The role of V2G is an active area of research. Because

---

(...continued)

"Environmental and Energy Implications of Plug-in Hybrid-Electric Vehicles," *Environmental Science & Technology*, Vol. 42 No. 4, 2008, pp.1185-1190.

[72] One study shows slightly less relative reductions in life cycle carbon emissions because the carbon emissions due to vehicle production (excluding the batteries) are similar, and battery production represents 2-5% of life cycle carbon emissions from a PHEV. See C. Samaras and K. Meisterling,"Life Cycle Assessment of Greenhouse Gas Emissions from Plug-in Hybrid Vehicles: Implications for Policy," *Environmental Science & Technology*, Vol. 42 No. 9, 2008, pp. 3170-3176.

[73] J. Pont, Full *Fuel Cycle Assessment: Well-to-Wheels Energy Inputs, Emissions, and Water Impacts*, California Energy Commission, 2007.

[74] The Electric Power Research Institute (EPRI) has projected that, although coal-fired power plants would provide much of the charging for PHEVs in 2030, most charging would nonetheless be from sources with pollution control equipment (new, existing, and retrofitted). Emissions of NO_x, SO_2, and VOCs were therefore projected to go down with PHEV penetration. See M. Duvall and E. Knipping, *Environmental Assessment of Plug-In Hybrid Electric Vehicles. Volume 2: United States Air Quality Analysis Based on AEO-2006 Assumptions for 2030*, Electric Power Research Institute, 2007.

[75] L. Gaines et al. (2007), "Sorting through the Many Total-Energy-Cycle Pathways Possible with Early Plug-In Hybrids," Electric Vehicle Symposium (EVS23), Anaheim, CA, December 2-5, 2007.

[76] EPRI used an air quality model to project that the PHEVs would reduce population exposure to ozone and particulate matter. Another study used a utility simulation model to project that PHEV charging in Colorado would come primarily from natural gas-fired power plants if there were no change in the electric generating fleet. This would lead to significant reductions in NO_x and VOC emissions in the Denver metro area, leading to modest improvements in ozone concentrations. See G.L. Brinkman et al. "Effects of Plug-In Hybrid Electric Vehicles on Ozone Concentrations in Colorado," *Environmental Science & Technology*, Vol. 44 No.16, 2010, pp. 6256-6262.

[77] This conclusion depends on the anticipated cycle life and cost of EV batteries. However, controlled charging (without V2G) is still a potentially significant source of flexibility, with the ability to raise the minimum load and avoid (continued...)

---

electric vehicles in any form have yet to achieve significant market penetration, assessing their potential as a source of grid flexibility is difficult. However, analysis has demonstrated potential system benefits of both controlled charging and V2G.[78]

Barriers to Deployment and Policies to Increase Vehicle Electrification

A primary barrier to deployment of electric vehicles is their cost and availability. After the discontinuation of commercially produced electric passenger vehicles in the early 1990s, and before the introduction of the Nissan Leaf and Chevrolet Volt in late 2010, there were no mass-produced electric passenger vehicles available in the United States.[79] The costs of the current generation of EVs and PHEVs are high—with recent prices for the Leaf and Volt about $35,000 and $39,000, respectively.[80] The earliest projected deployment of fuel cell vehicles is 2015. While the performance of battery technologies continues to improve, it is unclear when costs will reach the point needed for large scale adoption.[81]

There are a number of federal and state policies targeted towards increasing the use of battery electric and fuel cell vehicles. These include R&D efforts through the American Recovery and Reinvestment Act of 2009 (ARRA, discussed in detail in the corresponding technology chapter). ARRA also provides $2 billion toward grants for the manufacturing of advanced battery systems and electric vehicle components. These funds are intended to support domestic manufacturing of advanced lithium-ion batteries and hybrid electric systems and components. To incentivize adoption, ARRA supports tax credits for the purchase of PHEVs. A comprehensive summary of current federal and state incentives is provided at the Alternative Fuels & Advanced Vehicles Data Center.[82]

Other critical barriers include a lack of existing infrastructure for vehicle fueling and charging. For hydrogen fueled FCEVs or HEVs, entirely new infrastructure is needed for fuel production, transport and refueling. (These issues are discussed in more detail in the hydrogen chapter.) For EVs, lack of charging infrastructure, combined with limited range of pure electric vehicles

---

(...continued)

curtailment. For additional discussion of the impact of battery life and cycling on the value of V2G, see S.B. Peterson, J.F. Whitacre, and J. Apt, "The Economics of Using PHEV Battery Packs for Grid Storage," *Journal of Power Sources*, No. 195, 2010, pp. 2377-2384; and R. Sioshansi, R. and P. Denholm, "The Value of Plug-In Hybrid Electric Vehicles as Grid Resources," *The Energy Journal,* Vol. 31 No. 3, 2010, pp. 1-23.

[78] Short, W., and P. Denholm. (2006) "A Preliminary Assessment of Plug-In Hybrid Electric Vehicles on Wind Energy Markets" NREL/TP-620-39729.

[79] Some non-highway, low speed vehicles (neighborhood electric vehicles, or NEVs) were available over this time.

[80] Prices are manufacturer's suggested retail price (MSRP) before federal tax incentives. Nissan USA, "Nissan LEAF," Web page, 2011. http://www.nissanusa.com/leaf-electric-car/index#/leaf-electric-car/index; General Motors, "2011 Volt," 2011. http://www.chevrolet.com/volt-electric-car/.

[81] General Motors, in its recent IPO stated "In some cases, the technologies that we plan to employ, such as hydrogen fuel cells and advanced battery technology, are not yet commercially practical and depend on significant future technological advances by us and by suppliers. For example, we have announced that we intend to produce by November 2010 the Chevrolet Volt, an electric car, which requires battery technology that has not yet proven to be commercially viable. There can be no assurance that these advances will occur in a timely or feasible way." Securities and Exchange Commission (2010) Amendment to No. 9 to Form S-1 Registration Statement under the Securities Act of 1933 General Motors Company http://www.sec.gov/Archives/edgar/data/1467858/000119312510262471/ds1a.htm#rom45833_2

[82] U.S. Department of Energy, Alternative Fuels and Advanced Data Center, "Federal & State Incentives and Laws," website, 2011. http://www.afdc.energy.gov/afdc/laws/state.

---

presents a barrier to large-scale adoption, especially for those who do not have access to secure charging at home.[83] Current electric rate structures also create a barrier, preventing both maximum benefit of controlled charging to the grid, and lowest-cost charging for the consumer.

---

[83] For more discussion of charging infrastructure issues see T. Markel, *Plug-in Electric Vehicle Infrastructure: A Foundation for Electrified Transportation,* Report No. CP-540-47951, National Renewable Energy Laboratory, 2010.

Chapter 4: Batteries for Grid Applications

Overview

Batteries are devices that store energy chemically. This report focuses on "secondary" batteries, which must be charged before use and which can be discharged and recharged (cycled) many times before the end of their useful life. For electric power grid applications, there are four main battery types of interest:

- Lead-acid

- High temperature "sodium-beta"

- Liquid electrolyte "flow" batteries

- Other emerging chemistries[84]

Lead-acid batteries have been used for more than a century in grid applications and in conventional vehicles for starting, lighting, and ignition (SLI). They continue to be the technology of choice for vehicle SLI applications due to their low cost. Consequently, they are manufactured on a mass scale. In 2010, approximately 120 million lead-acid batteries were shipped in North America alone.[85] Lead-acid batteries are commonly used by utilities to serve as uninterruptible power supplies in substations, and have been used at utility scale in several demonstration projects to provide grid support.[86] Use of lead acid batteries for grid applications is limited by relatively short cycle life. R&D efforts are focused on improved cycle-life, which could result in greater use in utility-scale applications.

Sodium-beta batteries include sodium-sulfur (NaS) units, first developed in the 1960s,[87] and commercially available from a single vendor (NGK Insulators, Ltd.) in Japan with over 270 MW deployed worldwide.[88] A NaS battery was first deployed in the United States in 2002.[89] There are now a number of U.S. demonstration projects, including several listed in **Table 3**. The focus of NaS deployments in the United States has been in electric distribution deferral projects, acting to reduce peak demand on distribution systems, but they also can serve multiple grid support

---

[84] Several of the battery types discussed in this chapter have been demonstrated or proposed for transportation applications as well. However, they also have challenges in achieving the energy density or other characteristics needed for storing large amounts of energy in mobile applications. Batteries for electric vehicles are discussed in Chapter 5.

[85] Battery Council International, "Breakdown of North American Battery Shipments (2001-2010)," Chicago, November 4, 2011. http://www.batterycouncil.org/LeadAcidBatteries/BreakdownofShipments/tabid/173/Default.aspx

[86] Electric Power Research Institute and U.S. Department of Energy (EPRI/DOE), *EPRI-DOE Handbook of Energy Storage for Transmission and Distribution Applications*, Palo Alto, CA, No. 1001834, December, 2003. A 10 MW, 40 MWh lead-acid battery was built in Southern California in 1988. It operated for about nine years. A 21 MW, 14 MWh lead-acid plant was built in Puerto Rico in 1994 to provide spinning reserves. It operated for about five years.

[87] X. Lu et. al., "Advanced materials for sodium-beta alumina batteries: status, challenges, and perspectives," *Journal of Power Sources*, No. 195, 2010, pp. 2431-2442.

[88] D. Rastler, "New Demand for Energy Storage," Electric Perspectives, Edison Electric Institute, September 2008.

[89] Nourai, A., "Installation of the First Distributed Energy Storage System (DESS) at American Electric Power (AEP): A Study for the DOE Energy Storage Systems Program." SAND2007-3580. Albuquerque, NM: Sandia National Laboratories, June 2007. http://www.electricitystorage.org/images/uploads/docs/Sandia_First_Storage_AEP.pdf

---

services. An alternative high-temperature battery, sodium-nickel-chloride, is in the early stages of commercialization.[90]

Table 3. Example NaS Battery Installations in the United States

Size[a] (MW/MWh)	Developer	Location	Installation Date
0.1/0.72	AEP[b]	Gahanna, OH (First U.S. demonstration)	2002
1.0/7.2	AEP[b]	North Charleston, WV	2006
2.0/14.4	AEP[c]	Bluffton, OH; Balls Gap, WV; East Busco, IN	2008
1.0	New York Power Authority[d]	Long Island, NY	2008
4.0	AEP[c]	Presidio, TX	2009
1.0/7.2	Xcel Energy[e]	Luverne, MN	2009

Source: National Renewable Energy Laboratory compilation.

a. Continuous rating.

b. A. Nourai, *Installation of the First Distributed Energy Storage System (DESS) at American Electric Power (AEP): A Study for the DOE Energy Storage Systems Program*, SAND2007-3580, Sandia National Laboratories, Albuquerque, NM, June, 2007.

c. AEP, *Energy Storage in T&D Applications*, slide presentation, May 2009. http://www.aeptechcentral.com/docs/NAS-Presentation.pdf

d. G. Sliker, "Long Island Bus: NaS Battery Energy Storage Project," slide presentation, New York Power Authority, September 29, 2009. http://www.sandia.gov/ess/docs/pr_conferences/2008/sliker_nypa.pdf

e. Xcel Energy, "Wind-To-Battery Project," fact sheet, November 2008. http://www.xcelenergy.com/staticfiles/xe/Corporate/Environment/wind-to-battery%20fact%20sheet.pdf

"Flow" batteries, in which a liquid electrolyte flows through a chemical cell to produce electricity, are in the early stages of commercialization. In grid applications there has been some deployment of two types of flow battery: vanadium redox and zinc-bromide. There are a number of international installations of vanadium redox units, including a 250 kW installation in the United States to relieve a congested transmission line.[91] There are also a number of zinc-bromine demonstration projects.[92] Several other flow battery chemistries have been pursued or are under development, but are less mature.

In addition to the three battery types discussed above, there are several emerging technologies based on new battery chemistries which may also have potential in grid applications. Several of these emerging technologies are being supported by DOE efforts such as ARPA-E and are discussed briefly in the R&D section of this chapter.

---

[90] J. Baker, "New Technology and Possible Advances in Energy Storage," *Energy Policy*, Vol. 36, 2008, pp. 4368–4373.

[91] EPRI/DOE, 2003. The U.S. unit was installed by Pacificorp in 2004 in Moab, UT.

[92] EPRI/DOE, 2003.

Technology

Description and Performance

Lead-Acid

The lead-acid battery consists of a lead dioxide positive electrode (cathode), a lead negative electrode (anode), and an aqueous sulfuric acid electrolyte which carries the charge between the two. During discharge, each electrode is converted to lead sulfate, consuming sulfuric acid from the electrolyte. When recharging, the lead sulfate is converted back to sulfuric acid, leaving a layer of lead dioxide on the cathode and pure lead on the anode. In such conventional "wet" (flooded) cells, water in the electrolyte is broken down to hydrogen and oxygen during the charging process. In a vented wet cell design, these gases escape into the atmosphere, requiring the occasional addition of water to the system. In sealed wet cell designs, the loss of these gases is prevented and their conversion back to water is possible, reducing maintenance requirements. However, if the battery is overcharged or charged too quickly, the rate of gas generation can surpass that of water recombination, which can cause an explosion.

In "valve regulated gel" designs, silica is added to the electrolyte to cause it to gel. In "absorbed glass mat" designs, the electrolyte is suspended in a fiberglass mat. The latter are sometimes referred to as "dry" because the fiberglass mat is not completely saturated with acid and there is no excess liquid. Both designs operate under slight constant pressure. Both also eliminate the risk of electrolyte leakage and offer improved safety by using valves to regulate internal pressure due to gas build up, but at significantly higher cost than wet cells described above.[93]

Lead-acid is currently the lowest-cost battery chemistry on a dollar-per-kWh basis. However, it also has relatively low specific energy (energy per unit mass) on the order of 35 Wh/kg and relatively poor "cycle life," which is the number of charge-discharge cycles it can provide before its capacity falls too far below a certain percentage (e.g., 80%) of its initial capacity. While the low energy density of lead-acid will likely limit its use in transportation applications, increase in cycle life could make lead-acid cost-effective in grid applications.

The cycle life of lead-acid batteries is highly dependent on both the rate and depth of discharge due to corrosion and material shedding off of electrode plates inside the battery. High depth of discharge (DoD) operation intensifies both issues. At 100% DoD (discharging the battery completely) cycle life can be less than 100 full cycles for some lead-acid technologies. During high rate, partial state-of-charge operation, lead sulfate accumulation on the anode can be the primary cause of degradation. These processes are also sensitive to high temperature, where the rule of thumb is to reduce battery life by half for every 8°C (14°F) increase in temperature above ambient.[94] Manufacturers' warrantees provide some indication of minimum performance expectations, with service life of three to five years for deep cycle batteries, designed to be mostly discharged time after time. SLI batteries in cars have expected service lives of five to seven years, with up to 30 discharges per year depending on the rate of discharge. Temperature also affects

---

[93] D. Linden and T. Reddy, *Handbook of Batteries*, 3rd ed., McGraw Hill, New York, 2002.
[94] Ibid.

capacity, with a battery at -4°C (25°F) having between roughly 70% and 80% of the capacity of a battery at 24°C (75°F).[95]

For many applications of lead-acid batteries, including SLI and uninterruptible power supply (UPS), efficiency of the batteries is relatively unimportant. One estimate for the DC-DC (direct current) efficiency of utility-scale lead acid battery is 81%, and AC-AC (alternating current) efficiency of 70%-72%.[96]

High Temperature Sodium-Beta

Sodium-beta batteries use molten (liquid) sodium for the anode, with sodium ions transporting the electric charge. The two main types of sodium-beta batteries are distinguished by the type of cathode they use. The sodium-sulfur (Na-S) type employs a liquid sulfur cathode, while the sodium-nickel chloride (Na-NiCl2) type employs a solid metal chloride cathode. Both types include a beta-alumina solid electrolyte material separating the cathode and anode. This ceramic material offers ionic conductivity similar to that of typical aqueous electrolytes, but only at high temperature. Consequently, sodium-beta batteries ordinarily must operate at temperatures around 300°C (572°F).[97] The impermeability of the solid electrolyte to liquid electrodes and its minimal electrical conductivity eliminates self discharge and allows high efficiency.[98]

Technical challenges associated with sodium-beta battery chemistry generally stem from the high temperature requirements. To maintain a 300°C operating point the battery must have insulation and active heating. If it is not maintained at such a temperature, the resulting freeze-thaw cycles and thermal expansion can lead to mechanical stresses, damaging seals and other cell components, including the electrolyte.[99] The fragile nature of the electrolyte is also a concern, particularly for Na-S cells. In the event of damage to the solid electrolyte, a breach could allow the two liquid electrodes to mix, possibly causing an explosion and fire.[100]

Na-S batteries are manufactured commercially for a variety of grid services ranging from short-term rapid discharge services to long-term energy management services.[101] The DC-DC efficiency is about 85%. Calculation of the AC-AC efficiency is complicated by the need for additional heating. The standby heat loss for each 50 kW module is between 2.2 and 3.4 kW. As a result of this heat loss, plus losses in the power conversion equipment, the AC-AC efficiency for load-leveling services is estimated in the range of 75%-80%.[102] Expected service life is 15 years at 90% DoD and 4500 cycles.[103]

---

[95] EPRI/DOE, 2003.

[96] This estimate is of the Chino 10 MW battery with 96% inverter efficiency. EPRI/DOE, 2003.

[97] X. Lu et al., "Advanced Materials for Sodium-Beta Alumina Batteries: Status, Challenges, and Perspectives," *Journal of Power Sources*, No. 195, 2010, pp. 2431-2442.

[98] D. Linden and T. Reddy, 2002.

[99] Ibid.

[100] X. Lu et. al, 2010.

[101] B. Norris, J. Newmiller, and G. Peek, *NAS Battery Demonstration at American Electric Power*, SAND2006-6740, Sandia National Laboratories, 2007. NGK sells a "PS" module rated for continuous discharge for load-leveling applications and a "PQ" module rated for short discharge applications such as frequency and contingency reserves.

[102] 75% from Nourai, 2007 and 80% from A. Nourai, V.I. Kogan, and C.M. Schafer,"Load Leveling Reduces T&D Line Losses," *IEEE Transactions on Power Delivery*, Vol. 23, No. 4, 2008, pp. 2168–2173.

[103] EPRI/DOE, 2003.

The primary sodium-beta alternative to the Na-S chemistry, the Na-NiCl2 cell (typically called the ZEBRA cell).[104] Although ZEBRA batteries have been under development for over 20 years, they are only in the early stages of commercialization.[105] Nickel chloride cathodes offer several potential advantages including higher operating voltage, increased operational temperature range (due in part to the lower melting point of the secondary electrolyte), a slightly less corrosive cathode, and somewhat safer cell construction, since handling of metallic sodium—which is potentially explosive—can be avoided.[106] They are likely to offer a slightly reduced energy density.[107]

Liquid Electrolyte Flow Batteries

Flow batteries use liquid electrolytes that are pumped through a "stack" which contains either an ion-exchange membrane or an electrode array. Energy is stored primarily in active materials dissolved into electrolytes, which are stored externally and passed through the electrodes during charge and discharge. The electrodes are separated by an ion exchange membrane to keep the cathode-side and anode-side electrolytes separate. The advantage of flow battery technology is that the power component (MW) and the energy component (MWh) can be sized independently, with the electrolyte materials held in large external storage tanks for multi-MW applications. The power rating of a flow battery is determined by the size of the battery stack, and the energy rating by the size of the electrolyte storage tanks.

As stated earlier in this section, the two main types of flow battery in early commercialization are vanadium redox and zinc-bromine. Vanadium redox batteries are part of a large class of flow batteries using an ion-exchange membrane similar to that used in fuel cells. (Hence, they are sometimes called regenerative fuel cells.) In a redox flow battery, the active materials are always dissolved in the electrolyte. While a number of electrolyte materials have been proposed or are under development, vanadium has the greatest degree of commercialization, with a number of installations and active vendors. Other redox flow battery chemistries have yet to be commercialized but have the potential to provide cost-effective alternatives and are discussed in the R&D section of this chapter.

The redox flow battery offers several benefits over conventional batteries. First, the amount of energy storage available is limited only by the size of the tanks and the amount of electrolyte available. An additional benefit is avoiding the need to correct for differences among individual battery cells (cell balancing) typical in multi-cell storage configurations using other battery technologies, which allows for relatively simple construction of higher voltage redox batteries. Redox flow batteries can also be recharged mechanically by replacing the electrolyte. The disadvantages generally stem from the complexity of electrolyte pumping and storage; control system complexity; and relatively low specific energy and energy density (typically less than that of lead acid cells). The use of an ion-exchange membrane introduces other challenges and benefits. Leakage across the membrane is possible, causing mixing of the cathode-side and anode-side electrolytes. In a vanadium redox battery, the impact of leakage is mitigated by the

---

[104] The name derives from the Zeolite Battery Research Africa Project which invented the technology in 1985.

[105] See, for example, Daimler AG, "The New Mercedes-Benz A-Class E-CELL," web page, September 15, 2010, http://media.daimler.com/dcmedia/0-921-941776-1-1331063-1-0-0-0-0-0-11702-614316-0-1-0-0-0-0-0.html; J.L. Sudworth, "The Sodium/Nickel Chloride (ZEBRA) Battery," *Journal of Power Sources*, Vol. 100, 2001, pp. 149-163.

[106] C. Dustmann, "Advances in ZEBRA Batteries," *Journal of Power Sources*, Vol. 127, 2004, pp. 85-92.

[107] D. Linden and T. Reddy, 2002.

---

fact that both electrolyte materials are identical. However, using alternative chemistries, mixing can seriously degrade performance. The challenge of cost-effectively manufacturing reliable ion-exchange membranes is cited as a primary reason for the limited development of some of the earliest flow battery chemistries. The claimed calendar lifetime for a vanadium battery stack is at least 10 years with more than 10,000 cycles.[108]

The primary alternative to redox flow batteries is a flow battery where at least one of the active materials is plated onto an electrode. Several chemistries have been investigated, with zinc bromine being the most well developed. During charging, zinc is plated onto the negative electrode. When discharging zinc is dissolved into the electrolyte. This configuration benefits from a low-cost electrolyte and a slightly improved energy density, but problems can arise with formation of sharp particulates during zinc plating. A 2003 estimate of cycle life is about 2000 cycles or 6000 hours of continuous operations.[109] Commercial units have cycle life ratings from 1500 to over 2000 cycles.[110]

The DC-DC round-trip efficiency for flow batteries is in the range of 70%-80%.[111] However, as with all batteries, DC-AC losses reduce this efficiency further, and flow batteries have additional parasitic loads of electrolyte pumps. As a result, estimates of AC-AC roundtrip efficiency is in the range of 65%-72%.

Cost

The ability to estimate the capital cost of batteries varies by commercial maturity and application. Lead acid batteries are the most mature and lowest-cost technology with one 2008 estimate in the range of $150-$200/kWh.[112] To this must be added the costs of a storage installation equipment in addition to the battery cell itself (balance of plant), which were estimated at $265/kW in 2003; however, more recent estimates are considerably higher.[113] On a cost basis alone, this makes lead-acid batteries appear competitive for a wide variety of applications. However, this total cost must be placed in context of the relatively short cycle life of current lead-acid technology, restricting its use to applications which require few actual cycles per year.

One 2009 estimate for the cost of NaS battery is about $350-$400/kWh and about $450-$550/kW for the balance of plant.[114] This corresponds to about $2970-$3450/kW for a 7.2 hour device. It is

---

[108] EPRI/DOE, 2003.

[109] EPRI/DOE, 2003. Data from current vendors indicate longer lives are possible with periodic maintenance. Since zinc-bromine, like all flow battery technologies are in the early stages of commercialization, additional field trials will be necessary to establish calendar and cycle life estimates.

[110] P. de Boer, and J. Raadschelders, "Flow Batteries," white paper prepared for Leanardo ENERGY, June 2007, http://www.leonardo-energy.org/webfm_send/164; G. P. Corey, "An Assessment of the State of the Zinc-Bromine Battery Development Effort," RedFlow Limited, Brisbane, Australia, October 2010, http://www.redflow.com.au/docs/assessment_zinc_bromine_battery.pdf.

[111] EPRI/DOE, 2003; D. Rastler,"New Demand for Energy Storage," *Electric Perspectives*, September/October 2008 http://www.eei.org/magazine/EEI Electric Perspectives Article Listing/2008-09-01-EnergyStorage.pdf.

[112] D. Ton, et al. "Solar Energy Grid Integration Systems – Energy Storage (SEGIS-ES)," U.S. Department of Energy and Sandia National Laboratories, May 2008, http://www1.eere.energy.gov/solar/pdfs/segis-es_concept_paper.pdf.

[113] EPRI/DOE, 2003.

[114] D. Rastler, "Overview of Electric Energy Storage Options for the Electric Enterprise," slide presentation, Electric Power Research Institute, Palo Alto, CA, 2009, http://www.greentechmedia.com/images/wysiwyg/News/EPRIEnergyStorageOverview%20DanRastler.pdf

---

unclear whether or not this estimate considers potential deployment at scale. Another estimate from the first large NaS project in the United States claims that when initial project costs are removed, NaS would cost about $2500/kW for a 7.2 hour device (**Figure 5**).[115]

Figure 5. Cost Components for an Installed NaS System

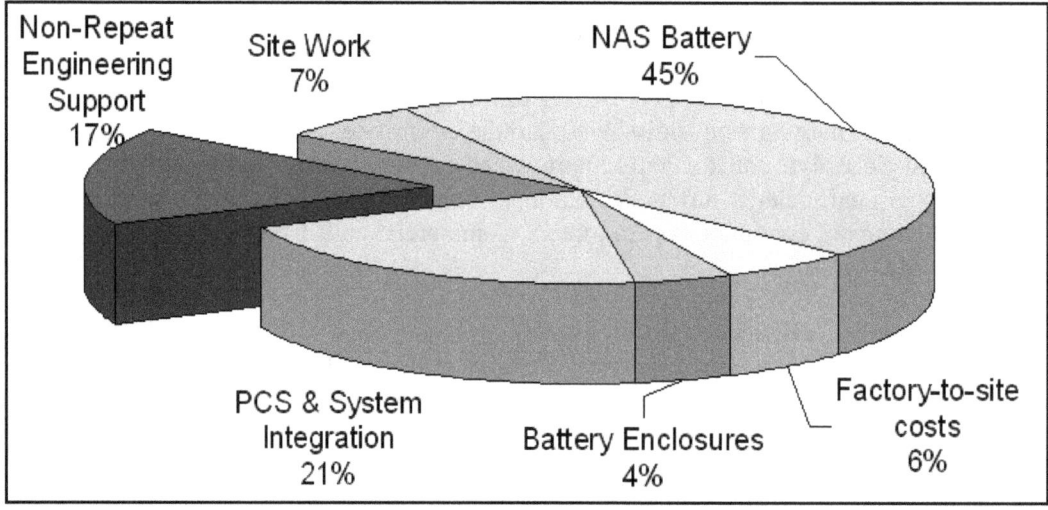

Source: A. Nourai, 2007.

Notes: PCS=Power Control Station. The factory-to-site shipping costs are considerable: "The total transportation costs from factory to site, including customs and handling charges plus a few other items shipped by air, translated to approximately $140/kW."

Limited cost estimates are available for both vanadium and zinc-bromine batteries. Lack of recent large-scale installations also makes cost estimates highly uncertain. A 2004 estimate for the cost of a vanadium flow battery is $236/kWh and $566 for the balance of plant, or about $2450/kW for an 8-hour device.[116] While total costs are not provided, a 2010 estimate for a proposed vanadium plant provides a cost breakdown of 35% electrolyte, 9% membrane, 17% other stack, and 5% power control station, with the remaining 34% for engineering, management, and balance of plant.[117]

A 2003 estimate[118] for Zinc-Bromine is $353/kWh and $576 for the balance of plant, or about $3400/kW for an eight hour device, while a 2011 manufacturer's estimate is about $780/kWh for the entire system, with projected costs of about $400/kWh for a next generation system at "full production levels."[119] More recent, unpublished estimates place flow battery costs in excess of $4000/kW for multi-hour devices, while a 2009 EPRI estimate places the *projected* costs of a generic flow battery at $1545-$3100/kW for a 4 hour device.[120] Manufacturing and deployment at

---

[115] A. Nourai, 2007.

[116] Electric Power Research Institute and U.S. Department of Energy (EPRI/DOE*), EPRI-DOE Handbook Supplement of Energy Storage for Grid Connected Wind Generation Applications,* No. 1008703, December 2004.

[117] J.F Startari, "Painesville Municipal Power Vanadium Redox Battery Demonstration Project," slide presentation, Ashlawn Energy, Painesville, OH, 2010. http://www.sandia.gov/ess/docs/pr_conferences/2010/startari_ashlawn.pdf

[118] EPRI/DOE, 2003.

[119] The manufacturer also projects future costs at "grid scale production" of close to $100/kWh. ZBB 2011 "Investor Presentation," http://www.zbbenergy.com/investor-relations/presentations/.

[120] D. Rastler, 2009.

scale will be necessary to establish better estimates of flow battery costs. Deployment at scale will also be needed to determine longevity as well as operation and maintenance requirements.

Some perspective on the overall cost reduction potential for certain battery types is provided in one recent analysis of different battery chemistries.[121] **Figure 7** shows the cost of various chemical pairs (couple elements) for battery types considered in this chapter. For many types, such as the NaS battery the cost of raw materials is, theoretically, a trivial component. Most others have couple element costs of about $10/kWh or less (assuming costs in the study year). One chemistry that stands out as a potential cost challenge is vanadium, with the couple element costs close to $100/kWh, primarily due to the high cost of vanadium. **Figure 6** also shows goals for the Department of Energy's ARPA-E grid storage and electric vehicle (EV) programs. The ARPA-E goal of $100/kWh appears to include both the power and energy component, including power conditioning equipment, installation, and other balance of system components. This would correspond to $800/kW for a device with eight hours of storage capacity, which would require battery costs of well below $100/kWh considering balance of system is currently a considerable fraction of $800/kW. The goal for the EV battery pack is discussed in the next chapter.[122]

Figure 6. Extraction Costs of Elements in Grid Battery Couples

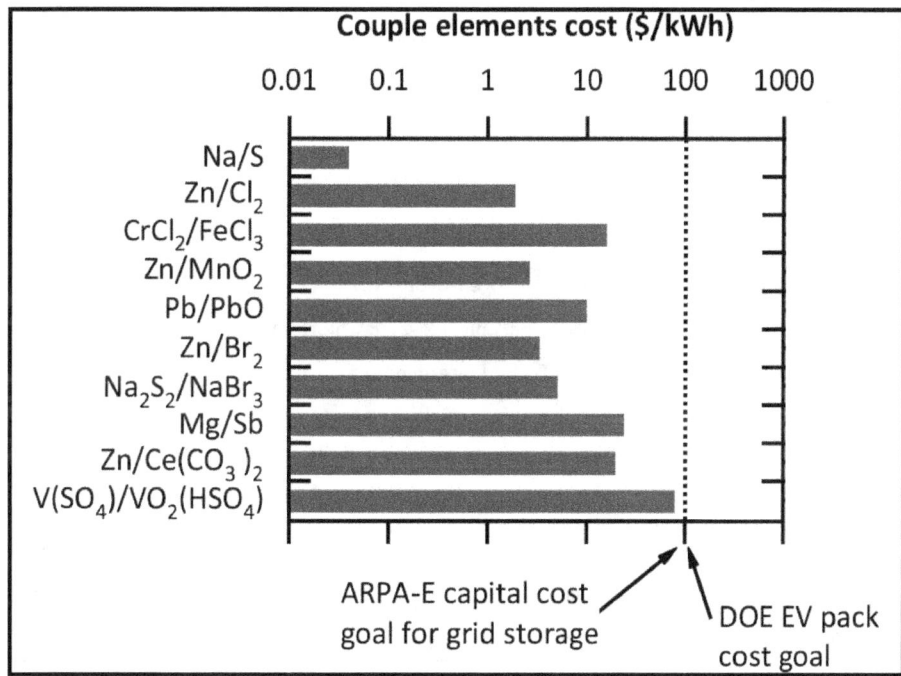

Source: C. Wadia, P. Albertus, and V. Srinivasan, 2011.

Notes: Calculated from U.S. Geological Survey element prices. The EV battery pack goal of $100/kWh includes only the cost of the battery itself.

---

[121] C. Wadia, P. Albertus, and V. Srinivasan, "Resource Constraints on the Battery Energy Storage Potential for Grid and Transportation Applications," *Journal of Power Sources*, Vol. 196, 2011, pp. 1593-1598.

[122] U.S. Department of Energy, *Grid-Scale Rampable Intermittent Dispatchable Storage (GRIDS)*, DE-FOA-0000290, CFDA Number 81.135, poncept paper, April 23, 2010.

Research and Development

The federal government, industry consortiums, and other groups support R&D efforts across a range of battery technologies, including a number of emerging battery chemistries that do not fall under the categories summarized above. The latter include alkaline, sodium ion, and liquid metal batteries. To illustrate these efforts, **Tables 4 and 5** list federal battery R&D activities supported by ARPA-E and ARRA, according to general battery type and chemistry. Activities associated with specific battery types are discussed below. Note that these tables include several lithium-ion and metal-air batteries, both of which are thought of as prime candidates for transportation applications. (These two battery types are discussed in greater detail in the following chapter.) However, some forms may be more suitable for grid applications and supported through grid-oriented R&D efforts.

In addition to federally supported efforts, there are grid battery technology R&D activities funded by other groups and among private companies (whose details may be proprietary). For example, the Stanford University's Global Climate and Energy Project has awarded grants to outside researchers for new grid-oriented battery technologies including enhanced electrolyte and solid oxide flow battery systems.[123]

Table 4. ARPA-E Supported Activities on Grid Battery Storage in FY2010-2011

Lead Research Organization	Battery Type /Chemistry	Funding ($Millions)
CUNY Energy Institute	Other (Zinc-Manganese Oxide)	3.00
Fluidic Energy, Inc.	Metal Air (Zinc Air)	3.00
General Atomics	Lead-Acid/Flow	1.99
Lawrence Berkeley National Lab	Flow (Hydrogen-Bromine)	1.59
Primus Power	Flow (Zinc Chloride-Zinc Chloride)	2.00
United Technologies Research Center	Flow (To Be Determined)	3.00
Univ. of Southern California	Metal Air (Iron-Air)	1.46
Arizona State University	Metal Air (Zinc-Air)	5.13
EaglePicher Technologies	Sodium-Beta (Sodium Sulfur)	7.20
Envia Systems	Lithium-Ion	4.00
Inorganic Specialists, Inc.	Lithium-Ion	2.00
Massachusetts Institute of Technology	Other (Liquid Metal)	6.95

Source: Sandia National Laboratories, "ARPA-E Awarded Projects in Energy Storage," web page, 2010, http://www.sandia.gov/ess/docs/ARPA-E_FY10-11_combined.pdf.

---

[123] Mark Shwartz, "GCEP Awards $3.5 million for Energy Storage Research," *Stanford Report*, Stanford University, September 23, 2011.

Table 5. ARRA Supported Grid Battery Demonstrations

Awardee	Battery Type /Chemistry	Size (Power/Energy)	Funding ($Millions)
Duke Energy Business Services	TBD	24 MW	$21.81
Primus Power	Flow (Zinc-Chloride)	25 MW (75 MWh)	$14.00
Southern California Edison Co.	Lithium-Ion	8 MW (4 hrs)	$24.98
City of Painesville	Flow (Vanadium Redox)	1 MW (6-8 MWh)	$4.24
Detroit Edison	Lithium-Ion	25 kW (20 units, 50 kWh each)	$5.00
East Penn Manufacturing Co.	Lead-Acid (with ultracapacitor)	3 MW (1-4 MWh)	$2.54
Premium Power Corp.	Flow (Zinc-Bromine)	5-500 kW (6 hrs)	$6.06
Public Service Company of NM	Lead-Acid	500kW (2.5MWhr)	$2.51
Aquion Energy, Inc.	Other (Sodium-Ion)	10-100 kWh	$5.18
Ktech Corp.	Flow (Iron-Chromium)	250kW (1MWhr)	$4.76
Seeo, Inc.	Lithium-Ion	25 kWh	$6.20

Source: Sandia National Laboratories, "ARRA Energy Storage Demonstrations," October 13, 2010, http://www.sandia.gov/ess/docs/ARRA_StorDemos_10-13-10.pdf.

Lead-Acid

The primary disadvantages of lead-acid batteries are their poor energy density and short cycle life. Marginal gains to specific energy can be achieved by improving the active material and design of the electrodes, but will always be limited by the chemistry's relatively low theoretical boundaries. Cycle life potentially can be increased by adding carbon in various forms to either the anode or cathode,[124] or by replacing the traditional lead acid anode with a carbon anode similar to that of an asymmetrical electrochemical capacitor.[125] Another approach to improve cycle life is the so called lead acid flow battery, in which lead is dissolved in an aqueous methanesulfonic acid electrolyte. This system differs from traditional flow batteries by using of just one electrolyte and the subsequent lack of troublesome electrolyte separators.[126] If long deep discharge cycle life is proven and costs can be kept low, these technologies may be promising for grid-based bulk electricity storage applications.

Sodium-Beta

There are several R&D efforts associated with sodium-beta batteries. One is to develop a stacked planar cell design that could cut cell costs in half.[127] This departure from the traditional tubular

---

[124] P.T. Moseley et al., "The Role of Carbon in Valve-Regulated Lead-Acid Battery Technology," *Journal of Power Sources*, Vol. 157, 2006, pp. 3-10; Enos, D., Hund, T., Shane, R. (2010) "Carbon-Enhanced VRLA Batteries." DOE Energy Storage Systems Research Program Annual Peer Review. http://www.sandia.gov/ess/docs/pr_conferences/2010/enos_snl.pdf

[125] L.T. Lam and R. Louey, "Development of ultra-battery for hybrid-electric vehicle applications," J. Power Sources 158: 1140-1148 (2006); P.T. Moseley et al (2006).

[126] Hazza, et. al., "A Novel Flow Battery: A Lead Acid Battery Based on an Electrolyte with Soluble Lead(II). Part I: Preliminary Studies," *Phys. Chem. Chem. Phys.*, 2004 (6) 1773-1778.

[127] Pacific Northwest National Laboratory, "EaglePicher Teams with PNNL to Transform Large-Scale Energy (continued...)

design has the ability to increase specific energy and power (the latter a limiting factor for the use of these batteries in many applications), improve packing efficiency, and improve modularity. It also presents the opportunity to address long term corrosion problems. However, planar designs face sealing and material selection challenges.[128] Other R&D efforts focus on low temperature sodium based chemistries using new cathodes and/or sodium ion conductors.[129] While cost is a major R&D focus, longevity and reliability still have room for some marginal improvement using improved cell configurations and designs.

Liquid Electrolyte

Flow battery R&D efforts include improving the performance of commercially available products and developing new chemistries. For vanadium redox cells, research seeks to decrease the vanadium required and increase energy density, for example, by up to 70%.[130] New redox couples that increase efficiency, improve specific energy, or utilize more cost effective or less toxic materials are also the subject of investigation. These chemistries include iron-chromium, zinc-chloride, hydrogen-halogen, hydrogen-bromine, lead, and others.[131] An earlier flow battery type (sodium-bromide/sodium-polysulfide) reached the initial stages of commercialization, but was discontinued. It is unclear if development of this chemistry is being pursued.[132] Another key requirement for large-scale deployment will be achieving demonstrated reliability and longevity. These requirements are complicated by the toxic and corrosive electrolytes, which pose significant materials challenges for the hydraulic subsystems and ion exchange membrane, particularly for chemistries other than vanadium in which electrolyte mixing is unacceptable.

Other Emerging Technologies

Slight modifications to the ubiquitous alkaline battery (e.g., Duracell batteries) utilizing a powdered zinc anode, manganese dioxide cathode, and potassium hydroxide electrolyte, make the system rechargeable. These batteries are currently plagued by very short and highly DoD dependent cycle lives (on the order of 10 or fewer cycles at high DoD) with excessive capacity loss between cycles (approaching 50% between the first and second cycle in some cases).[133] Due

(...continued)

Storage," web page, PNNL-SA-72347. http://energyenvironment.pnl.gov/highlights/highlight_4.asp

[128] X. Lu et. al., "Advanced Materials for Sodium-Beta Alumina Batteries: Status, Challenges, and Perspectives," *Journal of Power Sources*, Vol. 195, 2010, pp. 2431-2442.

[129] C-W Park et. al., "Room-Temperature Solid-State Sodium/Sulfur Batteries," *Electrochemical and Solid State Letters*, Vol. 9, No. 3., 2006, pp. A123-A125.

[130] Liyu Li et al., "A New Vanadium Redox Flow Battery Using Mixed Acid Electrolytes," Presentation to the U.S. DOE Energy Storage Systems (ESS) Program Review, November 2, 2010. http://www.sandia.gov/ess/docs/pr_conferences/2010/li_pnnl.pdf

[131] T.M Anderson and ID. ngersoll, "MetILs: New Ionic Liquids for Flow Batteries," Presentation to the U.S. DOE Energy Storage Systems (ESS) Program Review, November 2, 2010, http://www.sandia.gov/ess/docs/pr_conferences/2010/anderson_snl.pdf. Iron chromium was the first flow battery chemistries explored, but development was limited by the state of ion exchange membranes.

[132] "Company Pulls Plug on Power Storage Plant in Lowndes County," *Associated Press*, December 9, 2003; EPRI/DOE, 2004; This chemistry, originally trademarked as Regenesys was nearly commercialized with two partially constructed facilities, including one in the United States They were cancelled when the parent company discontinued development for "business reasons."

[133] D. Linden and T. Reddy, 2002.

to this massive life cycle performance differential between this chemistry and others, rechargeable alkaline batteries have only a small and declining market share. However, modifications have been proposed to overcome cycling challenges, with one ARPA-E supported "high-risk" project.[134]

Sodium ion batteries function in principle like lithium-ion batteries do—by shuttling positively charged ions between electrodes. Sodium ion batteries differ from the sodium beta batteries by the use of non-reactive electrode materials allowing the elimination of the ceramic separator and enabling room temperature operation.[135] The combination of highly available materials with aqueous electrolyte and low voltage cells has the potential to provide the low cost and high safety necessary for grid applications, with cost claims competitive with lead-acid, but with cycle life exceeding 5000 cycles and 100% DoD.[136] There are a number of other emerging chemistries in the early stages of research, including nitrogen-air[137] and a liquid metal battery.[138]

Deployment Challenges

Batteries for utility applications have yet to be deployed at scale, so potential challenges, notably land use, are not well quantified. One estimate for land use of a lead-acid facility is 77 m^2/MW for a device with 15 minutes of capacity.[139] For NaS, an estimate is about 211 m^2/MW (with a 7.2 hour storage capacity).[140] An estimate for a proposed (and subsequently cancelled) 12 MW, 100-120 MWh flow battery is about 850 m^2/MW[141] with additional land surrounding the facility.[142]

Availability of raw materials is another major concern for some battery types, especially those also being considered for large-scale transportation applications. However, this factor appears to be less critical for most of the battery types being considered for grid applications discussed in this chapter.[143] **Figure 7** provides an estimate of the 2010 production and reserves of battery materials, measured in Terawatt-hours (TWh) of energy storage potential (ESP). For additional context, the installation of 100 GW of storage, with 10 hours of capacity (equivalent to about 10% of the total installed generation capacity in the United States) would require about 1 TWh of

---

[134] S. Banerjee et al., "Flow-Assisted Rechargeable Zn-MnO2 Battery," Poster, U.S. DOE Energy Storage Systems (ESS) Program Review, November 2, 2010. http://www.sandia.gov/ess/docs/pr_conferences/2010/slide6_banerjee_cunyei.pdf

[135] Liu, J., "Emerging Technologies for Large-scale Energy Storage: Towards Low Temperature Sodium Batteries," Pacific Northwest National Laboratory, 2006. Presentation to the U.S. DOE Energy Storage Systems (ESS) Program Review, November 2, 2010. http://www.sandia.gov/ess/docs/pr_conferences/2010/liu_pnnl_2.pdf

[136] T. Wiley, "Demonstration of a Sodium Ion Battery for Grid Level Applications," Aquion Energy, Presentation to the U.S. DOE Energy Storage Systems (ESS) Program Review, November 2, 2010. http://www.sandia.gov/ess/docs/pr_conferences/2010/wiley_aquion.pdf

[137] F. Delnick, D. Ingersoll, and K. Waldrip, "Nitrogen-Air Battery," Sandia National Laboratories Presentation to the U.S. DOE Energy Storage Systems (ESS) Program Review, November 2, 2010. http://www.sandia.gov/ess/docs/pr_conferences/2010/ingersoll_snl.pdf

[138] M. LaMonica, "Liquid Metal Battery Snags Funding from Gates Firm," *CNET News*, May 20, 2011. http://news.cnet.com/8301-11128_3-20064404-54.html

[139] EPRI/DOE, 2003.

[140] NGK Insulators, Ltd., "Principle of the NAS Battery," web page, October 10, 2011. http://www.ngk.co.jp/english/products/power/nas/principle/index.html

[141] EPR/DOE 2003.

[142] Tennessee Valley Authority, *Environmental Assessment: The Regenesys Energy Storage System*, August 2001.

[143] C. Wadia, P. Albertus, and V. Srinivasan, 2011.

---

storage capacity. While production rates might need to be increased if deployed at large scale, few of the materials appear to be a major limiting factor, with the exception of vanadium, where resource limits could present challenges at extremely large scale. Antimony (Sb) is even more constrained but a high-temperature Mg/Sb battery has not been demonstrated.

Figure 7. Energy Storage Potential (ESP) of Battery Material Reserves

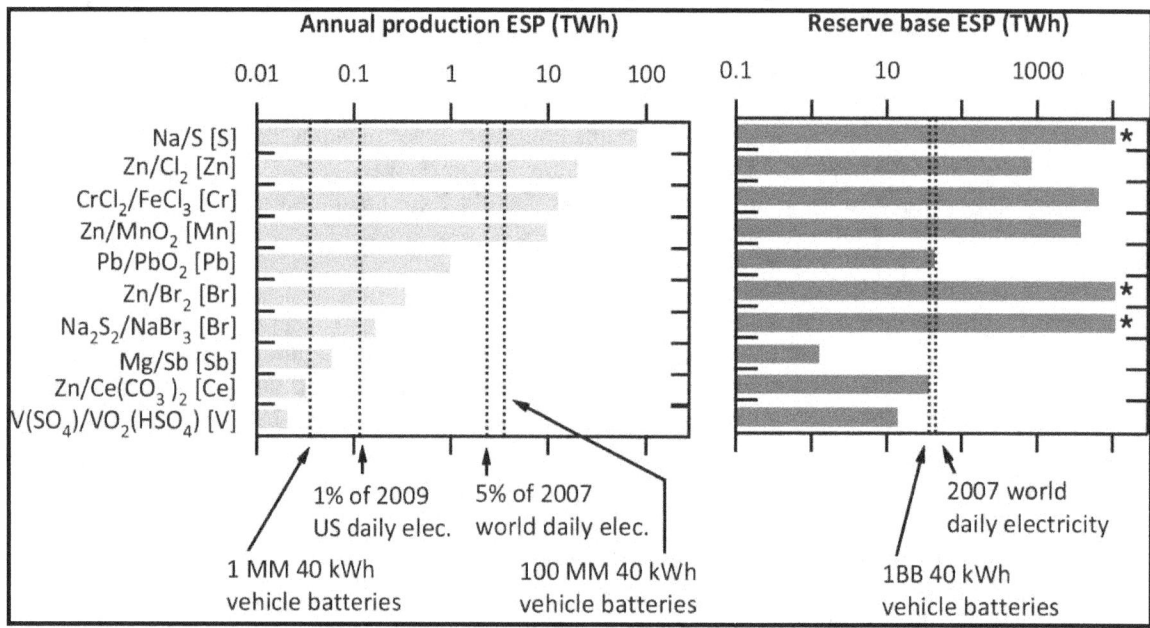

Source: C. Wadia, P. Albertus, and V. Srinivasan, 2011.

Notes: The elements in brackets at the right side of the labels are the limiting elements in each couple. The asterisk (*) indicates ESP well beyond the limit of the figure. Production is raw material for all uses.

Batteries under consideration use materials with a range of toxicity. Lead, for example, can be harmful to human health, and, thus, requires appropriate collection and recycling efforts to minimize potential health impact. For flow batteries, proper containment and mitigation is needed to address the potential release of materials from large tanks used for multi-MW applications.[144]

Conclusions

There are a large number of diverse battery chemistries in various stages of development and commercialization. Several projects have demonstrated competitive or near competitive economics for power grid applications. The rapid response of batteries makes them well suited for ancillary service applications such as frequency regulation, although they must demonstrate long calendar and cycle life, which is a challenge for many available battery technologies. Batteries will compete with other newly commercialized technologies such as flywheels for short-duration ancillary services. For longer-duration application, reduced capital cost is the primary requirement. A single battery technology has yet to emerge as a likely market leader for the many potential applications for grid storage. R&D will likely further improve battery technical performance and reduce costs for multiple technologies. Engineering and improved

---

[144] Tennessee Valley Authority, August 2001.

manufacturing techniques will also reduce costs and increase reliability for many of the battery types under development.

Chapter 5: Batteries for Electric Transportation

Overview

This chapter discusses battery technologies with the greatest potential for use in electric transportation. Compared to more grid-oriented storage technologies, batteries for electric vehicles need to have higher energy density, storing more energy for a given weight or volume, and therefore can be lighter, offering longer drive ranges. Because some vehicle battery technologies may also be suited for grid applications, the distinction may not be a rigid one, but it highlights significant differences in specifications and development that ease the discussion. For electric vehicle applications, there are four main battery types of interest:

- Nickel-based aqueous

- Lithium-ion

- Lithium metal

- Metal-air

Nickel-cadmium and nickel-metal hydride are the two main nickel-based aqueous (liquid electrolyte) battery types. Nickel-cadmium batteries have been challenged by the cost and toxicity of cadmium (they have largely been banned in the European Union), and are not considered a viable technology for large-scale deployment in vehicles. Nickel-metal hydride (NiMH) batteries are deployed extensively in current hybrid electric vehicles, such as the Ford Escape and Chevrolet Malibu hybrid models, and are, therefore, included in this chapter. In these applications, the battery serves primarily as a power resource for vehicle starting and acceleration. However, due to low energy density compared to lithium-ion or emerging battery technologies, current NiMH technology is unlikely to see large-scale deployment for plug-in hybrid electric vehicles (PHEVs) or battery electric vehicles (EVs).

Figure 8 shows the growth of hybrid electric (HEV) passenger vehicle sales over the last decade. Altogether, 1.7 million HEVs were sold in the United States between 2003 and 2010.[145] Most of these vehicles used NiMH batteries; only recently have HEVs and EVs begun to use lithium-ion batteries.[146]

---

[145] Thomas B. Reddy, Editor, *Linden's Handbook of Batteries, Fourth Edition*, McGraw-Hill, 2011, p. 29.26; For classification by manufacturer and other details see Electric Drive Transportation Association, "Hybrid Vehicle Sales Information," web page, June 2010. http://www.electricdrive.org/index.php?ht=d/Articles/cat_id/5514/pid/2549

[146] The first hybrid to transition to lithium-ion was the Mercedes-Benz S-Class sedan using a Johnson Controls-SAFT (JCS) battery. K. Snyder, X.G. Yanand T.J. Miller, "Hybrid Vehicle Battery Technology—The Transition from NiMH to Li-Ion," SAE Technical Paper No. 2009-01-1385, Society of Automotive Engineers, Warrendale, PA (2009).

Figure 8. United States Hybrid Electric Vehicle Sales

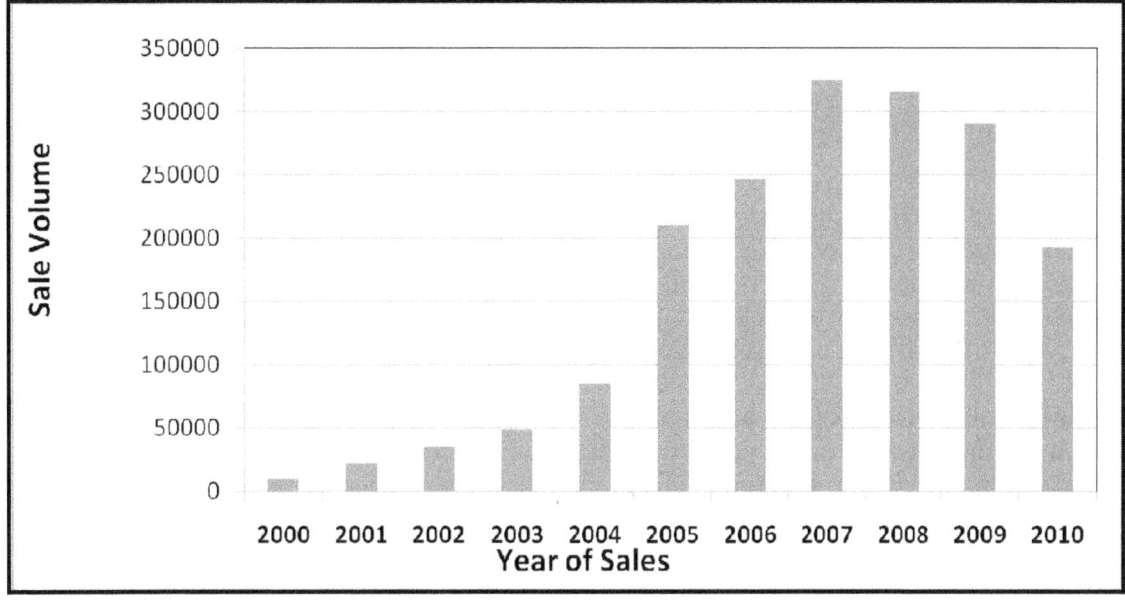

Source: Reddy, 2011.

Lithium-ion (Li-ion) batteries are by far the most popular battery type for portable consumer electronics, due primarily to their durability, high specific energy, correspondingly light weight, and reasonably fast-charge/discharge capability.[147] Recently, lithium-ion batteries have begun to enter the automotive market in hybrid and full electric vehicles. Several manufacturers including SAFT, LG Chem, SK Energy, Hitachi, AESC, A123, Enerdel, and Panasonic have developed high-power lithium-ion HEV batteries now under various stages of testing and commercialization. In addition to the ability to store more energy per unit weight, the automotive market also benefits from this chemistry's high power, efficiency, and long cycle life capability.[148] However, lower costs and enhanced safety are required before Li-ion batteries significantly impact the transportation sector.[149] While the focus of lithium-ion battery development has been for mobile and transportation applications, they are also being deployed in grid applications. Examples include an 8 MW Li-ion battery system installed by AES to provide frequency regulation in New York, and a 2 MW installation in Southern California, with a number of projects proposed or under development.[150]

Two other battery technologies (lithium metal and metal-air) currently under development promise up to a tenfold increase in specific energy. Although these chemistries have demonstrated basic performance and energy density potential in niche applications, they are still in the R&D stage for deployment in the transportation sector.

---

[147] M. Armand and J.M., Tarascon, "Building Better Batteries," *Nature*, No. 451, February 7, 2008, pp. 652-657; M.S. Whittingham, "Lithium Batteries and Cathode Materials," Chemical Reviews, Vol. 104, No. 10, 2004, pp. 4271-4302.

[148] F. Wagner, B. Lakshmanan, and M. Mathias, "Electrochemistry and the Future of the Automobile," *Journal of Physical Chemistry Letters*, Vol. 1., No. 14, 2010, pp. 2204-2219.

[149] U.S. Department of Energy, Office of Energy Efficiency and Renewable Energy, *FreedomCAR and Fuel Partnership: 2009 Highlights of Technical Accomplishments*, 2009. http://www1.eere.energy.gov/vehiclesandfuels/pdfs/program/2009 fcfp accomplishments_rpt.pdf

[150] AES Energy Storage, "AES Energy Storage Projects," web page, 2011. http://www.aesenergystorage.com/projects.html

Technology

Description and Performance

Nickel

All nickel-based aqueous batteries use a nickel oxyhydroxide cathode and a potassium hydroxide (KOH) electrolyte. Differentiation among nickel battery chemistries is in the anode. The original anode material in commercial nickel batteries, cadmium, resulted in a battery with significantly improved energy density and cycle life over lead acid batteries. However, because cadmium is toxic, further deployment of nickel-cadmium technology will be limited.

In NiMH cells, a complex metal alloy (comprised of some rare-earth metals, nickel, zirconium, and aluminum) is used to store hydrogen on the negative electrode. This chemistry has higher specific energy, cycle life, and high discharge rate capability than nickel-cadmium cells, but suffers from poor low temperature performance (below 0°C/32°F) and limited shelf-life—as low as three months.

The use of metallic zinc as an anode increases cell voltage, capacity, and improves high rate performance. Furthermore, the relative abundance of zinc[151] keeps the cost lower than both nickel-cadmium and NiMH batteries. However, the nickel-zinc battery suffers the same drawbacks as other systems with a metallic anode. In such a system, metal is deposited back on the anode during charge, as a highly non-uniform layer, leading to the formation of sharp particulates (known as dendrites) and swelling of the battery. Large volume changes cause mechanical stress on other cell elements, leading to degraded performance. Notably, the dendrites can create internal short circuits and loss of active material, causing irreversible capacity loss. Iron has also been employed as an anode for nickel-based batteries, but currently, is in limited use due to problems with the formation of hydrogen gas, leading to pressure build-up within the cells.[152]

Other nickel-based batteries, such as nickel-hydrogen have been demonstrated. A nickel-hydrogen cell is essentially a hybrid battery-fuel cell, with gaseous, pressurized hydrogen used as the anode active material. Designed and employed exclusively for aerospace applications, these cells can provide exceptionally long life along with other benefits, but their extremely high cost essentially prevents their use in all other applications.[153]

Lithium-Ion

Lithium-ion (Li-ion) batteries operate by shuttling lithium ions to the anode structure when charging, then by migrating the same ions across a porous separator via the electrolyte to the

---

[151] U.S. Geological Survey, *Mineral Commodity Summaries 2011*, January 2011, p. 189.

[152] D. Linden T. Reddy, 2002.

[153] L. H. Thaller and A.H. Zimmerman, *Nickel-Hydrogen Life Cycle Testing: Review and Analysis*, Aerospace Press Series, Aerospace Press, CA, 2003.

cathode structure during discharge. There are multiple combinations of cathodes, anodes, and electrolytes commercially classified under the lithium-ion umbrella as discussed below.[154]

Most Li-ion battery development efforts over the past decade have been focused on cathode technology. The anodes have generally been composed of common carbon-based materials (e.g., graphite, hard carbon, etc.). The anodes also are relatively less expensive components in the cell, and are not the limiting factor in terms of battery energy. However, improvements in anode performance may still yield gains, and may affect safety risks from the plating out of metallic lithium at the anode, especially when a battery is fast-charged at low temperatures.[155]

Three classes of cathode material prevail today: layered transition-metal oxides (e.g., cobalt oxide and various mixed oxides such as nickel-cobalt-aluminum oxide, etc.), spinels (e.g., manganese oxide), and olivines (e.g., iron phosphate).[156] The main difference among these classes of cathodes is in their crystal structures. Each offers various advantages and disadvantages as shown in **Figure 9**. The largest drawbacks of the layered oxide cathodes are cost and safety. Cost is driven up by cobalt and nickel content. Safety concerns are driven by the production of oxygen under abusive conditions (typically high temperature, high voltage and state of charge, where reaction with the electrolyte or dissolution may occur). Since the electrolyte is comprised of flammable organic solvents, when combined with oxygen, such reactions lead to the risk of fire and explosion. The principal challenges in spinel-based cathodes are lower energy content compared to layered oxides and a tendency for manganese to dissolve at higher temperatures (>55 °C/131 °F), thereby limiting longevity.[157] Olivines offer significant improvements in the safety threshold and stability. However, the maximum voltage that olivine materials can offer is lower than the other two categories, and the energy content is roughly half of that in layered oxides.

Figure 9. Comparison of the Various Lithium-Ion Battery Chemistries

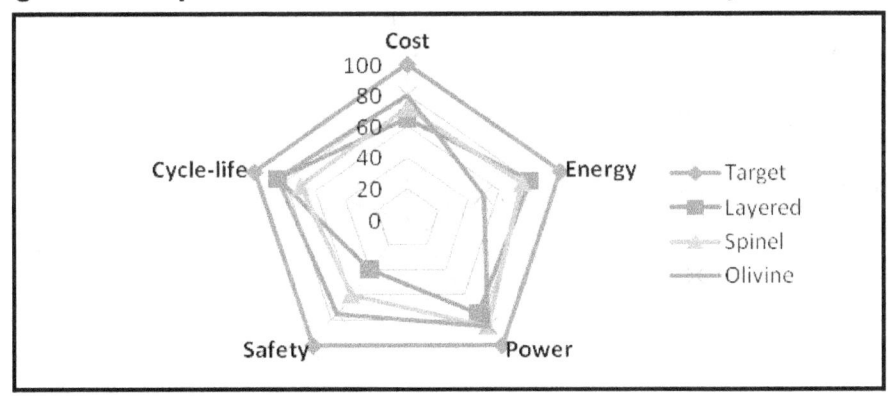

Source: S. Santhanagopalan, National Renewable Energy Laboratory, 2011.

Note: Scale is based on typical industry targets.

---

[154] G.-A. Nazri, G. Pistoia, Editors, *Lithium Batteries: Science and Technology*; Kluwer Academic Publishers, New York, 2004.

[155] N.A. Chernova, et al., "Layered Vanadium and Molybdenum Oxides: Batteries and Electrochromics," Journal of Materials Chemistry. No. 19, 2009, pp. 2526-2552; A.N. Jansen et. al., "Low-temperature Study of Lithium-ion Cells Using a LiySn Micro-Reference Electrode," Journal of Power Sources, No. 174, 2007, pp. 373-379.

[156] B.L. Ellis, K.T. Lee, and L.F. Nazar, "Positive Electrode Materials for Li-Ion and Li-Batteries," *Chemistry of Materials*, Vol. 22, 2010, pp. 691-714.

[157] G.-A. Nazri and G. Pistoia, 2004.

The electrolyte commonly consists of a mixture of organic solvents, such as an ethylene carbonate/dimethyl carbonate, or propylene carbonate mixtures that contain a dissolved fluoride based lithium salt. Due to the wide voltage window for lithium-ion batteries (4 volts compared to 2 volts or less for other batteries), the electrolyte is often exposed to working conditions beyond its stability limits. One way to overcome these stability issues—particularly on the carbon anodes—is to build a protective layer known as the solid electrolyte interface (SEI) layer. The SEI layer is typically formed during the first few charge cycles. Tuning the electrolyte formulation to optimize the SEI layer is an actively pursued means to improve the longevity of these batteries.

The cycle life of current Li-ion batteries is highly dependent on how deeply the battery is discharged each cycle. Using data from one published study, a graphite/nickelate Li-ion battery might last for 3000 cycles at 80% DoD, and 500,000 cycles at 4% DoD.[158]

Lithium Metal

The operational potential of using lithium metal as the anode could be significant since it offers a ten-fold increase in energy stored per unit weight. This idea has been successfully implemented in the non-rechargeable batteries (e.g., Energizer Ultimate Lithium AAA Cells). Extension of these efforts into rechargeable cells has been met with some success, limited primarily due to the highly reactive nature of lithium metal. The lithium-sulfur chemistry is among the most successful so far.

The basic lithium-sulfur cell is comprised of a liquid sulfur cathode, generally supported with a porous carbon framework, a liquid electrolyte, and a lithium metal anode. The working principle of the lithium sulfur cell closely resembles that of the sodium-sulfur cells (see Chapter 4).[159] Lithium sulfur cells offer the potential for extremely high energy density (theoretical energy content is ~2500 Wh/kg compared to 200 Wh/kg for the lithium-ion cells), due in part to the low molecular weight of sulfur. The low cost and high availability of sulfur also aid the cost effectiveness and sustainability of manufacturing these batteries.[160] However, because the electrical conductivity of sulfur is low, it is generally necessary to employ porous carbon supports, the additional weight of which lowers the energy content per unit weight or volume of the cell.[161] The chemistry also includes a natural mechanism to protect the cell from overcharge by diverting the current to a shuttle reaction in which sulfur is cycled back-and-forth between two valence states. However, tuning the shuttle for optimal performance is complex, and limits the life of the cell to 10 months or less, in some cases. In addition, the multiple intermediate compounds that are formed during the charge and discharge of the battery make stabilizing the cathode a difficult task. Often these intermediates are insoluble and block the porous network within the sulfur cathode.[162]

---

[158] J.C. Hall, et al., "Decay Processes and Life Predictions for Lithium Ion Satellite Cells," Paper AIAA 2006-4078, 4th International Energy Conversion Engineering Conference and Exhibit, San Diego, CA, Jun. 26-29, 2006.

[159] X. Ji, K. T. Lee, and L.F. Nazar, "A Highly Ordered Nanostructured Carbon-Sulphur Cathode for Lithium-Sulphur Batteries," *Nature Materials,* Vol. 8, No. 6, 2009, pp. 500-506.

[160] Wang et. al., "Sulfur Composite Cathode Materials for Rechargeable Lithium Batteries," *Advanced Functional Materials*, Vol. 13, 2003, pp. 487-492.

[161] X. Ji et. al., 2009.

[162] V.S. Kolosnitsyn and E.V. Karaseva, "Lithium-Sulfur Batteries: Problems and Solutions," *Russian Journal of Electrochemistry*, Vol. 44, No. 5, 2008, 506-509.

The use of lithium metal as an anode provides extremely high energy, but comes with many challenges to safety and long-term operation. Lithium metal is deposited (plated) during charge and dissolved when the battery is used. This plating process results in a significant volume change, with cycling, on the order of 300% (compared to approximately 10% for graphitic anodes), inducing mechanical stress on all components of the cell and exacerbating other failure mechanisms. Furthermore, this process can be highly non-uniform, leading to possible dendrite formation. As in lithium-ion batteries, the electrolyte employed in lithium-sulfur batteries is also flammable, leading to potential safety concerns. However, when combined with the sulfur cathode, the anode surface and newly formed dendrites are coated by protective layers of soluble polysulfide chains. Finally, metallic lithium is extremely reactive when exposed to water, presenting safety concerns if the cell container is breached.[163]

Metal-Air

Metal-air batteries have the potential to simultaneously be the highest energy density and lowest-cost energy storage solution for many applications. In a metal-air battery, oxygen available from the atmosphere serves as the cathode, in conjunction with a metallic anode. When the battery is discharging, oxygen is combined with the metallic anode to create metal oxides; when charging, these metal oxides are reduced to plate the metal back at the anode. Although several metals have been considered for such systems—including magnesium, iron, aluminum, zinc, and lithium—only the latter two have shown any affinity for electrical recharging.[164] There is also a variant of metal-air systems, "metal-water" batteries, in which the air cathode is replaced with water. Many of the issues are the same as those of metal-air systems, but the voltage is significantly reduced and applications are generally limited, thus they are omitted from this discussion.

Zinc-air systems have a theoretical specific energy greater than 3 kWh/kg, and thus present an opportunity to greatly surpass the energy carrying capability of lithium-ion cells. Furthermore, their use of zinc, a highly abundant and low-cost material, offers improved sustainability and cost effectiveness. The deployment challenges of zinc-air batteries include poor reversibility and resultant cycling problems due to metal plating, as well as evaporation of the aqueous electrolyte (when used in an open system).

Lithium-air (Li-air) batteries, with a theoretical specific energy exceeding 11 kWh/kg (excluding oxygen), are perhaps the most likely candidate battery to approach the energy density of fossil fuels. Hence , Li-air technology is of considerable interest to automotive and other mobile applications. However, after including accessory components, such as the porous carbon support layers, the realized specific energy is only about 30% of the theoretical value. Even at these levels, however, Li-air specific energy still surpasses today's state of the art Li-ion technology (0.25 kWh/kg) by an order of magnitude. As in the case of Li-ion batteries, the use of metallic lithium (and organic electrolytes) raises safety concerns for Li-air cells. The large thermodynamic loss between charge and discharge reactions also raises issues of reversibility.[165]

[163] Ibid.; G.-A. Nazri and G. Pistoia, 2004.

[164] D. Linden and T. Reddy, 2002.

[165] J. Zhang et. al., "Air Dehydration Membranes for Nonaqueous Lithium-Air Batteries," *Journal of the Electrochemical Society*, Vol. 157, No. 8, 2010, pp. A940-A946.

Cost

DOE estimates for the current generation of Li-ion battery packs for electric vehicles are between $800 and $1200/kWh, including the battery cell, integration, thermal management, and other system costs.[166] The battery cells alone are believed to account for only 65% of these costs, leaving a large share of costs in the other pack level systems.[167] The cost of the pack relative to typical vehicle costs is a significant issue, however. For example, the necessary pack size for a small- to mid-size electric sedan is on the order of 30 kWh, implying a battery cost on the order of $20,000, which exceeds the total cost of many vehicles in this class. A more detailed discussion of the per-cycle cost of Li-ion batteries is provided in Chapter 7.

Figure 10 shows the cost levels that batteries would need to achieve for the fuel cost savings over five years to offset the initial incremental cost according to one study.[168] The chart assumes a baseline grid electricity cost of 9 cents/kWh and fuel costs of $2.15/gallon (low case) and $4.30/gallon (high case). Energy requirements for various kinds of vehicles are represented by the power-to energy ratio: plug-in hybrids (PHEVs) with long ranges typically have large energy requirements, whereas hybrid electric vehicles (HEVs) have more demanding power requirements. If the fuel prices were at $2.15/gallon, the cost of batteries should follow the curve labeled "Projected battery costs" relatively closely (at least for shorter ranges) in order to be cost-neutral; however, for the case of future costs of fuel, the energy storage systems only need to reach the $500-$700/kWh range, according to this study.

[166] This is the pack, not the cells alone, and includes integration, thermal management and other costs. US DOE (2010) Batteries for Electric Energy Storage in Transportation (BEEST) DE_FOA-0000207 DFDA Number: 81.135 3/1/2010. More recent estimates are less than $800/kWh. Kanellos (2011) "Is Sodium the Future Formula for Energy Storage? Greentech Media. http://www.greentechmedia.com/articles/read/is-sodium-the-future-formula-for-energy-storage/

[167] Boston Consulting Group, "Batteries for Electric Cars: Challenges, Opportunities and the Outlook to 2020," January 2010, http://www.bcg.com/documents/file36615.pdf.

[168] T. Markel and A. Simpson, "Plug-in Hybrid Electric Vehicle Storage System Design," presentation at the Advanced Automotive Battery Conference, Baltimore, MD, May 17-19, 2006. http://www.nrel.gov/docs/fy06osti/40237.pdf

Figure 10. Battery Cost Requirements for a Five-Year Payback from Fuel Savings

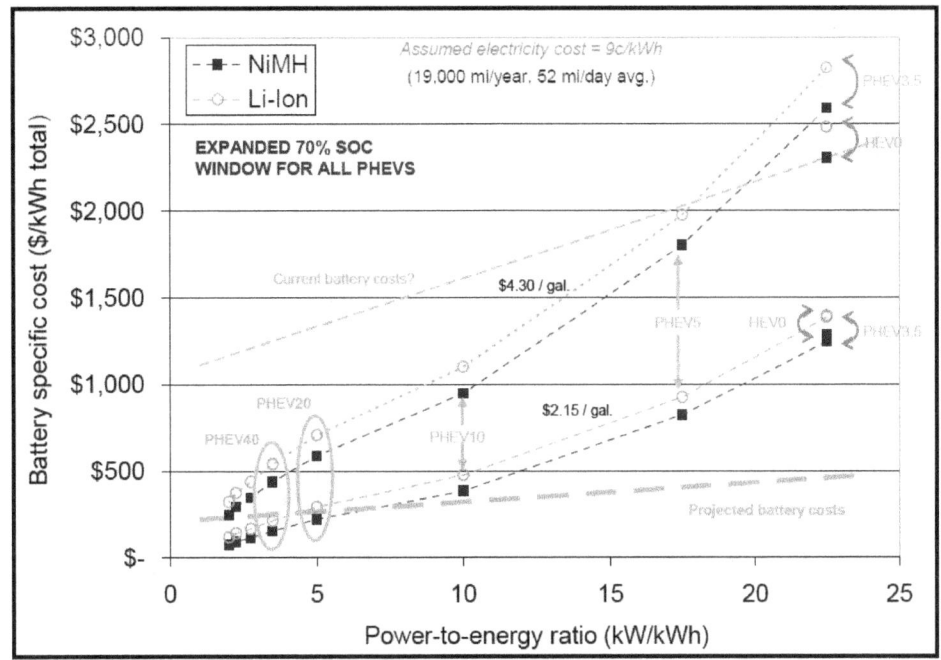

Source: T. Markel and A. Simpson, 2006.

Notes: This figure is for the entire battery pack, including the cells, integration, thermal management, etc. The number following "PHEV" in the figure indicates the miles that can be traveled solely using electricity ranging from 3.5 to 40.

The maturity of lithium-sulfur and metal-air battery systems is not yet sufficient to project expected battery pack costs. Cost challenges associated with raw materials (couple elements) for the battery technologies discussed in this chapter are illustrated in **Figure 11.** For several battery types, including nickel-based batteries and many Li-ion batteries, the couple element costs alone are a large fraction of $100/kWh, the ARPA-E and DOE cost goals for complete grid and vehicle storage battery packs, respectively.[169]

---

[169] Near-term goals are $500/kWh (2012) and $270-$300/kWh (2014) Longer-term goals EERE and ARPA-E goals are $100-$150/kWh. D. Howell, "Vehicle Technologies Program," presentation at the 2011 Annual Merit Review and Peer Evaluation Meeting, Energy Storage R&D, May 913, 2011, http://www1.eere.energy.gov/vehiclesandfuels/pdfs/merit_review_2011/electrochemical_storage/es000_howell_2011_o.pdf; see also U.S. Department of Energy, "Batteries for Electric Energy Storage in Transportation (BEEST)," DE_FOA-0000207, March 1, 2010.

Figure 11. Extraction Costs of Elements in Vehicle Battery Couples

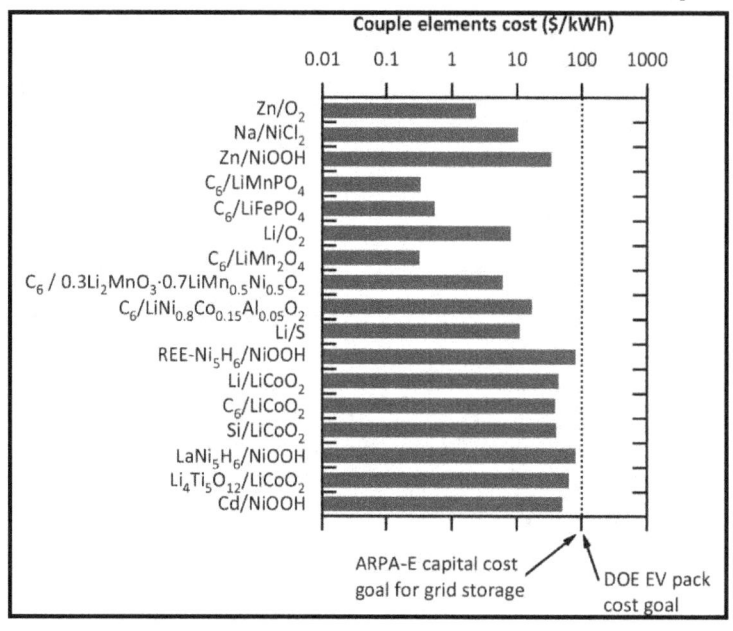

Source: C. Wadia, P. Albertus, and V. Srinivasan, 2011.

Notes: Calculated from U.S. Geological Survey element prices. The EV battery pack goal of $100/kWh includes only the cost of the battery itself.

Research and Development

At present, DOE R&D efforts on vehicle batteries are in two main programs: Batteries for Advanced Transportation Technologies, which is a fundamental research program focused on new materials, and Advanced Battery Research, which focuses on scale-up and commercialization of technologies. For incentivizing deployment at scale, ARRA has provided funding to a number of manufacturers, detailed in **Table 6**. The Department of Energy has also awarded grants for advanced battery R&D under the ARPA-E Batteries for Electrical Energy Storage in Transportation (BEEST) program, some of which are discussed below. In all, the BEEST program invested $40 million in non-conventional battery technologies in 2010.[170]

Private sector R&D efforts in the United States are conducted by the United States Advanced Battery Consortium (USABC), led by the "big three" U.S. auto makers—Chrysler, Ford, and General Motors—in cooperation with the DOE. USABC has funded a number of advanced battery development and technology assessment contracts, in some cases with DOE co-funding.[171]

---

[170] Advanced Research Projects Agency-Energy, "Batteries for Electrical Energy Storage in Transportation (BEEST)," web page, 2011. http://arpa-e.energy.gov/ProgramsProjects/BEEST.aspx

[171] United States Council for Automotive Research LLC, "USABC Awards $15.6 Million in Advanced Battery Technology Contracts to Three Firms," press release, March 2, 2011.

Table 6. ARRA Supported Vehicular Battery Demonstrations

Awardee	Funding Amount (M$)	Project Locations	Technology
Johnson Controls, Inc.	300	Holland, MI Lebanon, OR	Production of nickel-cobalt-metal battery cells and packs, as well as production of battery separators (by partner Entek) for hybrid and electric vehicles.
A123 Systems, Inc.	249	Romulus, MI Brownstown, MI	Manufacturing nano-iron phosphate cathode powder and electrode coatings; fabrication of battery cells and modules; and assembly of complete battery pack systems for hybrid and electric vehicles.
KD ABG MI, Llc.	161	Midland, MI	Production of manganese oxide cathode / graphite lithium-ion batteries for hybrid and electric vehicles.
Compact Power, Inc.	151	St. Clair, MI Pontiac, MI Holland, MI	Production of lithium-ion polymer battery cells for the GM Volt using a manganese-based cathode material and a proprietary separator.
EnerDel, Inc.	119	Indianapolis, IN	Production of lithium-ion cells and packs for hybrid and electric vehicles. Primary chemistries include manganese spinel/lithium titanate for high power applications; and manganese spinel/amorphous carbon for high energy applications.
General Motors Corp.	106	Brownstown, MI	Production of high-volume battery packs for the GM Volt. Cells will be from LG Chem, Ltd. and other cell providers to be named
Saft America, Inc.	100	Jacksonville, FL	Production of lithium-ion cells, modules, and battery packs for industrial and agricultural vehicles and defense applications. Primary lithium chemistries include nickel-cobalt-metal and iron phosphate
Exide Technologies with Axion Power International	34	Bristol, TN Columbus, GA	Production of advanced lead-acid batteries, using lead-carbon electrodes for micro and mild hybrid applications.
East Penn Manufacturing Co.	33	Lyon Station, PA	Production of the UltraBattery (lead-acid battery with a carbon supercapacitor combination) for micro and mild hybrid applications.

Source: Sandia National Laboratories, "Recovery Act Awards for Electric Drive Vehicle Battery and Component Manufacturing Initiative," 2010, http://www.sandia.gov/ess/docs/ARRA.pdf.

Longer-term targets for PHEV energy storage, as envisioned by the Department of Energy's FreedomCAR program, are shown in **Figure 12**. The figure highlights key barriers to achieving these long-term targets; all current battery chemistries face some or all of these barriers.

Figure 12. FreedomCAR PHEV Energy Storage Goals

	Short-Term	Long-Term
	SUV	Car
Discharge Power, kW	45	38
Regen Power, kW	30	25
Available Energy, kWh (Charge-Depleting)	3.4	11.6
Available Energy, Wh/kg	80-95	140-160
Available Energy, kWh (Charge-Sustaining)	0.5	0.3
Range, miles	10	40
Battery Mass, kg	60	120
Cold Cranking Power*, kW	7	
Cycle Life, Charge-Depleting Cycles	5,000	5,000
Calendar Life, years	10+	10+
Operating Temperature, °C	-30 to 52	
Selling Price**, $	1,700	3,400

* Three 2s pulses at -30°C with 10s rest between pulses **Price based on 100,000 batteries/year production level

Source: G. Henriksen, "Overview of Applied Battery Research," presentation to the Department of Energy Annual Merit Review, Washington, D.C., June 9, 2010, http://www1.eere.energy.gov/vehiclesandfuels/pdfs/ merit_review_2010/electrochemical_storage/es014_henriksen_2010_o.pdf.

Nickel

Further development of nickel-cadmium and nickel-hydrogen batteries is unlikely due to the use of toxic cadmium and high cost, respectively. R&D topics for NiMH batteries include improving cold temperature performance, reducing self-discharge rates, increasing power, and extending cycle life.[172] Cost is also an issue, but is difficult to address through R&D since economies of scale in production have already been achieved, and a large fraction of the cost of NiMH batteries is due to the cost of nickel. Nickel-zinc batteries offer improvements relative to NiMH, but they are currently limited by poor cycle life. Overcoming this obstacle requires a solution to the zinc dissolution and plating problems.

Lithium-Ion

A significant number of research efforts in industry, national laboratories, and academia are presently devoted to improving cost, safety, energy density, cold temperature performance, and longevity of Li-ion batteries—a key focus of DOE vehicle technology R&D. Especially for large format automotive cells, increasing the scale of manufacturing is often cited as a likely pathway to reduce cost. To this end, $2.4 billion in ARRA funds were awarded in late 2009 to create a U.S. manufacturing base capable of supporting the annual production of 500,000 electric vehicles by

---

[172] M.A. Fetcenko, et al., "Recent advances in NiMH battery technology," *Journal of Power Sources*, No. 165, 2007, pp. 544-551.

2015, resulting in a projected 70% decrease in battery cost.[173] A large portion of such scale-induced cost reductions is based on the commoditization of materials, reported to make up 60% of current cell costs.[174] The active materials alone (anode, cathode, and electrolyte) have been shown to make up approximately 20% of costs.[175] Thus, development of anodes and cathodes incorporating lower-cost materials (such as iron rather than cobalt) is another route actively being pursued.

Safety is the foremost concern for many current and potential Li-ion battery markets, especially in light of well publicized laptop computer fires and fires following Chevrolet Volt crash tests involving Li-ion battery technology. Although such events are isolated, they merit concern, particularly in automotive applications. Electrode coatings currently under investigation can stabilize the electrode-electrolyte interface, benefiting not only safety but also cell longevity.[176] Non-flammable electrolyte systems, still in need of further development, hold similar promise.[177] Finally, system level approaches have introduced better thermal management and protective circuitry.

Low-temperature response, long-term degradation, specific energy, and other aspects of battery performance also show room for improvement, although in many cases they are already superior to competing technologies. It should be noted that long-term improvements in cell degradation and specific energy, achievable with advanced cathodes and anodes, electrode coatings, and other technical enhancements, also have the potential to reduce the cost of Li-ion cells on a dollars per kWh basis. Noteworthy are efforts funded by ARPA-E, including three-dimensional electrodes developed by Applied Materials and all solid-state batteries developed by Planar Energy.[178]

Lithium titanate has been offered as an alternative to graphite as an anode material. Lithium titanate anodes may provide high stability by operating at a much higher voltage versus lithium than carbon anodes do, by greatly reducing the chance of lithium plating, and by eliminating electrolyte reduction and the need for the SEI layer. This chemistry improves safety, longevity, and efficiency, but has a significantly lower cell voltage. Combined with a specific capacity (capacity per unit mass) about half that of graphite, a lithium titanate cell's energy storage capability is reduced by as much as 50% compared to the conventional lithium-ion cells.[179] Similar concerns arise for other metal-oxide anodes currently under evaluation.[180] Using a silicon anode also is under extensive study, as it offers an extremely high theoretical specific capacity.

---

[173] U.S. Department of Energy, *Fiscal Year 2010 Annual Progress Report for Energy Storage R & D*, January 2011.

[174] B. Barnett, et al., "PHEV Battery Cost Assessment," presentation to the Department of Energy Annual Merit Review, Washington, D.C., TIAX LLC, 2009. http://www1.eere.energy.gov/vehiclesandfuels/pdfs/merit_review_2009/energy_storage/es_02_barnett.pdf

[175] U.S. Department of Energy, January 2011.

[176] Y.S. Jung, et al., "Enhanced Stability of LiCoO2 Cathodes in Lithium-Ion Batteries Using Surface Modification by Atomic Layer Deposition," *Journal of the Electrochemical Society 2010*, No. 157, pp. A75-A81; Y.S. Jung, et al., "Ultrathin Direct Atomic Layer Deposition on Composite Electrodes is Critical for Highly Durable and Safe Li-Ion Batteries," *Advanced Materials*, No. 22, 2010, pp. 2172-2176.

[177] L. Zinck et al., "Purification Process for an Inorganic Rechargeable Lithium Battery and New Safety Concepts," *Journal of Applied Electrochemistry*, Vol. 36, 2006, pp. 1291-1295.

[178] Advanced Research Projects Agency-Energy, 2011.

[179] G.-A. Nazri and G. Pistoia, 2004.

[180] S.-H. Lee, et al., "Reversible Lithium-Ion Insertion in Molybdenum Oxide Nanoparticles," *Advanced Materials*, No. 20, 2008, pp. 3627-3632; C. Ban, "Nanostructured Fe3O4/SWNT Electrode: Binder-Free and High-Rate Li-Ion Anode," *Advanced Energy Materials*, No. 22, 2010, pp. E145-E149.

Volume change in the cell during charge/discharge is the major concern. Such extreme volume expansion can cause particle fracturing and loss of electronic conductivity, leading to high irreversible capacity loss and vastly reduced cycle life.[181]

Lithium-Sulfur

The promise of practical specific energy that is at least twice that of Li-ion is enticing to long-term PHEV goals. However, much work remains to be done to improve Li-ion specific energy while addressing capacity decline, self discharge, and safety. There are many ongoing efforts to address these challenges, such as new cathode structures reliant on different porous carbons,[182] and possibly doped or functionalized porous carbons to stabilize the polysulfide products. Surface coatings for increased sulfur utilization, stability, and conductivity, as well as new electrolytes formulated for increased conductivity and shuttle control, etc. are also under investigation.[183] The ARPA-E BEEST program awarded $5 million last year to a consortium including Sion Power, LLC, BASF, Lawrence Berkeley National Laboratory, and Pacific Northwest National Laboratory to develop lithium-sulfur batteries with a 300 mile range between charges (about three times that of conventional lithium batteries).

Metal-Air

With some minor differences among the different design variants, the core challenges for metal-air batteries lie with the obstruction of the active electrode surface, along with other challenges typical of metallic anodes, such as volume change and loss of uniformity during cycling. Protecting and stabilizing the anode is of particular importance where lithium is used, as it not only represents a barrier to long-term performance, but also to safety. Most current R&D efforts focus on overcoming the energy loss between charge and discharge at the air cathode by employing suitable catalysts.[184] Additional challenges include evaporation of the electrolyte and contamination when using ambient air. The use of ionic and solid electrolytes has the potential to address evaporation, but typically degrades efficiency and increases cost. Alternatively, evaporation and air contamination can be approached at the system level by using closed or filtered air systems, but such systems bring added cost, complexity, and mass.

Noteworthy funding for metal-air battery R&D includes a $5 million grant under the ARPA-E program and support from the Oregon Department of Energy's Small Scale Energy Loan Program (SELP), to ReVolt Technology, LLC in 2010, that published datasheets demonstrating a few hundred cycles of charge-and-discharge on their Zinc-air batteries.[185]

---

[181] C. K. Chan, et al., "High-performance Lithium Battery Anodes Using Silicon Nanowires," *Nature Nanotechnology*, VCol. 3, 2008, pp. 31-35.

[182] B. L. Ellis, K.T Lee, and L.F Nazar, "Positive Electrode Materials for Li-Ion and Li-Batteries," *Chemistry of Materials*, Vol. 22, 2010, ,pp. 691–714.

[183] Y.-J. Choi, et al., "Effects of Carbon Coating on the Electrochemical Properties of Sulfur Cathode for Lithium/Sulfur Cell," *Journal of Power Sources*, Vol. 184, No. 2, 2008, pp. 548-552; D. Aurbach, et al., "On the Surface Chemical Aspects of Very High Energy Density, Rechargeable Li–Sulfur Batteries," *Journal of the Electrochemical Society*, Vol. 156. No. 8, 2009, pp. A694-A702.

[184] Y.-C. Lu et. al., "Platinum-Gold Nanoparticles: A Highly Active Bifunctional Electrocatalyst for Rechargeable Lithium-Air Batteries," *Journal of the American Chemical Society*, Vol. 132, No. 35, 2010, pp. 12170-12171.

[185] ReVolt Technology, "ReVolt Technology LLC Selected for $5 Million ARPA-E BEEST Grant Award," press release, May 5, 2010. http://www.revolttechnology.com/news.asp?id=28

Deployment Challenges

The primary deployment challenges of electric vehicle batteries are high initial cost and limited performance. Achieving large-scale deployment will require batteries with higher energy density and battery chemistries that use abundant materials. **Figure 13** provides the specific energy of a range of battery types compared to the DOE target for electric vehicle battery packs (200 Wh/kg). Currently, only three chemical couples can practically meet that goal: one variant of lithium-ion, one variant of lithium-metal, and one variant of metal-air (the upper right quadrant in **Figure 13**).

Figure 13. Practical vs. Theoretical Specific Energy for 27 Battery Chemistries

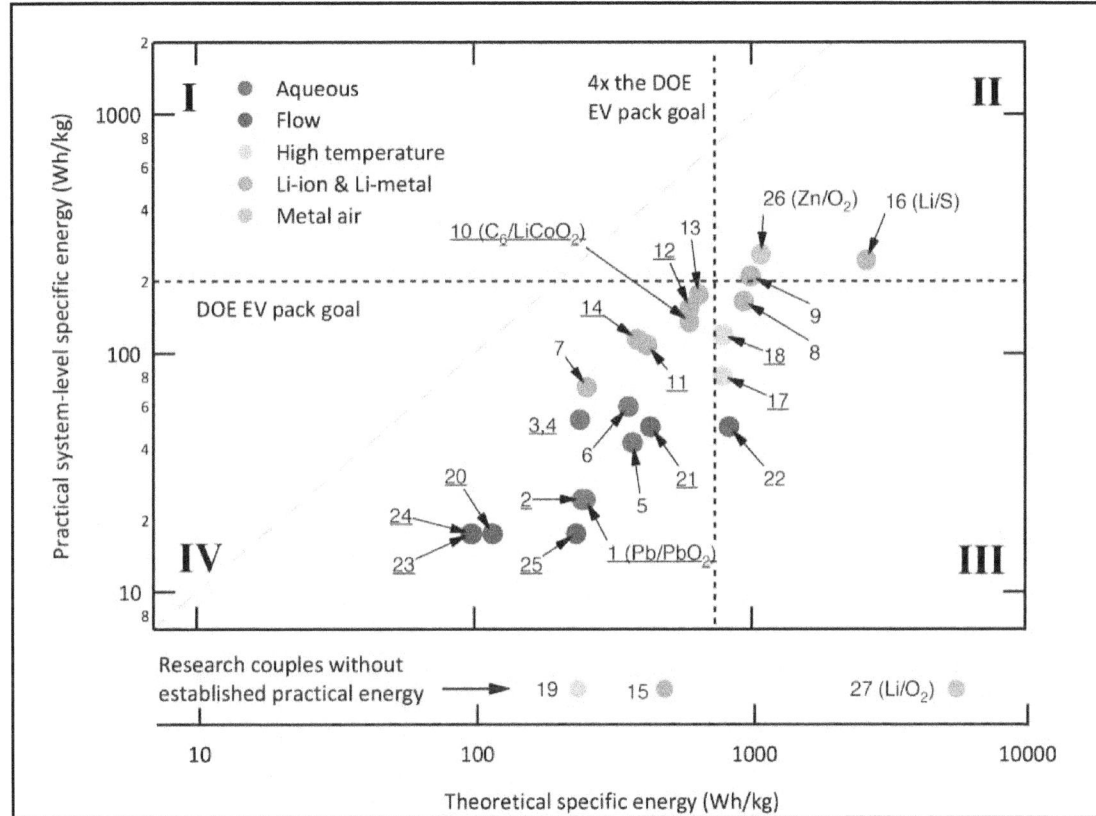

Source: C. Wadia, P. Albertus, and V. Srinivasan, 2011.

Key: 1= Lead-Acid, 2=Ni-Cd, 3-5=Nickel Metal Hydride, 7-16=Li-Ion and Li-Metal, 17=Sodium Nickel Chloride, 18=Sodium Sulfur, 20=Vanadium Redox, 21=Zinc Bromine, 22-25=Other Flow Batteries, 26-27=Metal-air.

Notes: Specific energy is based on the weight of active materials alone. The DOE pack goal for an EV with a 40 kWh battery pack is shown, as well as the approximate theoretical energy required for a couple to have a chance of meeting the pack goal. Chemistries that have demonstrated very good reversibility (i.e., a long cycle life) are underlined.

Another potential deployment challenge is the potential scarcity of raw materials. **Figure 14** provides an estimate of the 2010 production and reserves of materials for the batteries discussed in this chapter. The material requirements are measured in TWh, with total annual production (fro all uses) compared to the potential worldwide needs of 1 million and 100 million 40 kWh battery packs, and total estimate reserves compared to the need for 1 billion battery packs. This demonstrates that large-scale deployment of lithium-based batteries will require greatly increased production rates. Total material availability may also be a challenge for batteries using cobalt.

Figure 14. Energy Storage Potential (ESP) of Battery Material Reserves

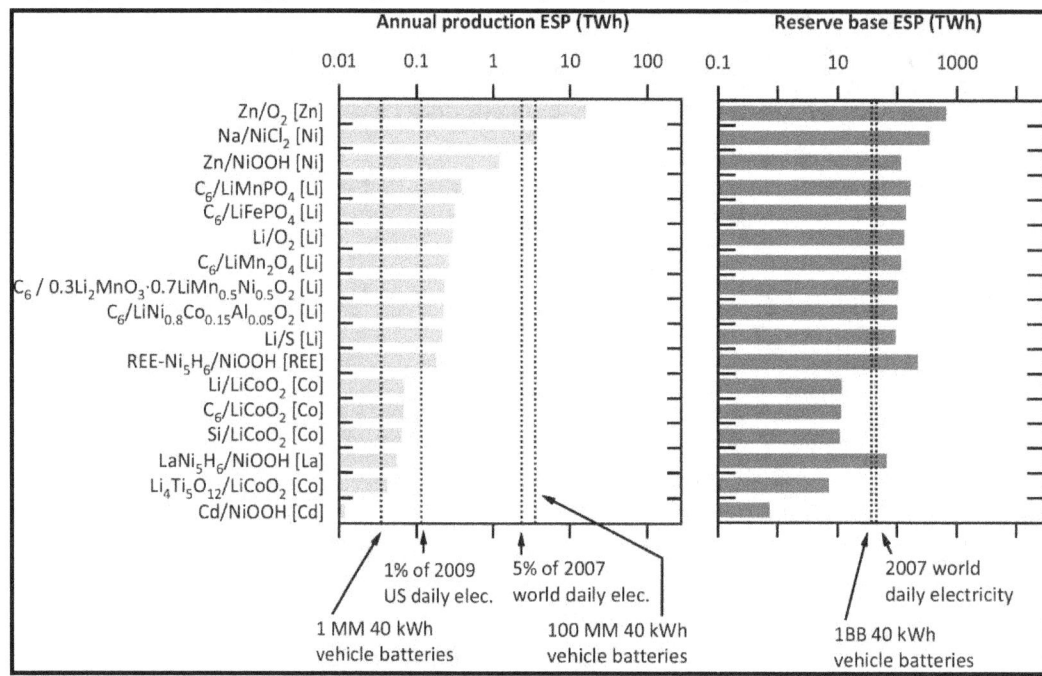

Source: C. Wadia, P. Albertus, and V. Srinivasan, 2011.

Notes: The elements in brackets at the right side of the labels are the limiting elements in each couple. Production is raw material for all uses

Conclusions

Batteries show potential as an option for electrifying the transportation sector. Reduced battery cost could produce PHEVs and EVs with life-cycle costs at or below those of conventional fossil-fueled vehicles, depending on the future price of gasoline. However, achieving this potential will require addressing the cost of a few key raw materials, engineering large format batteries to adequate safety standards, and improving long-term performance. A number of research efforts currently exist to address the shortcomings of current Li-ion technology, as well as to pursue advanced battery types that could provide even greater performance.

Chapter 6: Hydrogen

Overview

Hydrogen is a high quality "energy carrier" that can be used to store energy for use in vehicle or grid applications.[186] A hydrogen energy storage system would be composed of several components, with the specific configuration depending upon the application; it generally consists of hydrogen production (electrolysis), hydrogen transmission and storage, and hydrogen conversion to electricity (via fuel cell or combustion process). Hydrogen is typically discussed as being used in fuel cell systems, which have higher efficiency than conventional combustion engines, partly compensating for the upstream energy intensity of most hydrogen energy pathways. Fuel cells convert hydrogen and oxygen into electricity through an electrochemical process, resulting in water and heat as potentially useful byproducts. Fuel cells have been described as "open" or "flow" batteries, in which the chemical reactants, typically hydrogen and oxygen, are fed into the fuel cell continuously rather than being stored within the system like a battery.[187] Though the concept has been understood since 1838, the first significant use of fuel cells was in the Apollo space program in the 1960s.[188]

Hydrogen produced from natural gas is used extensively today to refine crude oil and produce fertilizer, but most of the R&D interest in hydrogen is for use as a sustainable fuel for transportation applications, such as buses, ground equipment (e.g., forklifts or airport tugs) and light duty vehicles. Hydrogen fueled vehicles would produce zero emissions at the point of use (if using fuel cells), with hydrogen being potentially produced from a variety of energy sources, illustrated in **Figure 15**. Hydrogen fuel cell electric vehicles (FCEVs) can potentially provide a cost effective, clean and low-carbon alternative to gasoline in light duty vehicle applications depending upon the original energy source for the fuel.[189] A 2010 study by McKinsey evaluating the current status of FCEVs concluded that the technology has moved from the demonstration phase to the commercial deployment phase.[190] Most major automotive companies have some demonstration-scale production FCEVs in operation. Hydrogen fuel cell systems are commercially viable today in forklifts for warehouses. Demonstration FCEV projects are ongoing in the United States and elsewhere, with commercial deployment in select areas expected in 2015

---

[186] Hydrogen is one of a number of gaseous or liquid fuels that can be produced by electricity for later use in a vehicle or stationary generator. Other chemical fuel pathways (such as ammonia) are possible, but most of the R&D on electricity based fuels is targeted towards hydrogen, which is the focus of this chapter. Hydrogen may also be produced from other primary fuels (e.g., natural gas) . Currently, the vast majority of hydrogen is produced by non-electric sources and processes.

[187] For additional information on fuel cell technology, see EG&G Technical Services, Inc., *Fuel Cell Handbook, 7th Edition*, prepared for the U.S. Department of Energy, November 2004. http://www.netl.doe.gov/technologies/coalpower/fuelcells/seca/pubs/FCHandbook7.pdf

[188] P. Hoffman, *The Forever Fuel: The Story of Hydrogen*, Boulder, CO, Westview Press, 1981.

[189] D.L. Greene, et al., *Hydrogen Scenario Analysis Summary Report: Analysis of the Transition to Hydrogen Fuel Cell Vehicles and the Potential Hydrogen Energy Infrastructure Requirements*, Oak Ridge National Laboratory, ORNL/TM-2008/030, March, 2008, http://info.ornl.gov/sites/publications/files/Pub10268.pdf; National Academy of Science, *Transitions to Alternative Transportation Technologies—A Focus on Hydrogen*, The National Academies Press 2008.

[190] McKinsey and Company, Inc., *A Portfolio of Power-Trains for Europe: A Fact-Based Analysis*, 2010. http://www.europeanclimate.org/documents/Power_trains_for_Europe.pdf

---

for ordinary consumers.[191] Projected costs for automotive fuel cell systems, once perceived to be a major cost barrier, have come down significantly due to R&D activities, with estimates approaching $50/kW when deployed at scale.[192] Significant industry and government stakeholder support for FCEVs has been realized internationally and in several states including California and Hawaii.[193] There are currently 56 hydrogen refueling stations in the United States, most being test or demonstration projects, and approximately 300 FCEVs are in use by select household consumers in California, refueling at some 22 hydrogen stations within the state.[194] Major barriers to large-scale deployment of FCEVs include the durability of fuel cells, sufficient onboard fuel storage, and the need for new refueling infrastructure.

For grid applications, challenges to hydrogen storage include capital costs and low round-trip efficiency (well under 50%) compared to other commercially available energy storage technologies. Fuel cell systems have become commercial in a number of niche markets, and electrolysis-based storage systems have been demonstrated in multiple countries. Current fuel cell applications include backup power, remote power, and combined heat and power for buildings, as well as space and military applications.[195] These various markets received approximately 24,000 fuel cell unit shipments in 2009, mostly for small portable applications, including over 1000 fuel cell forklifts and more than 600 backup power units at telecom sites. Of these shipments, 58 large grid fuel cell systems represented a total of about 15 MW of electricity production capacity.[196] Recent demonstrations storing and testing electrolytic hydrogen include a project at the National Renewable Energy Laboratory, in partnership with Xcel Energy, testing hydrogen production, storage, and conversion to grid electricity.[197] There are a number of other test projects of various scales internationally.[198] Japan has installed over 3000 small (1 kW) residential systems for combined heat and power as of 2008.[199]

---

[191] For international developments, see International Partnership for Hydrogen and Fuel Cells in the Economy (IPHE), website, 2011. http://www.iphe.net/

[192] S. Satyapal, "Hydrogen and Fuel Cells Technology Update," presented at the Fuel Cell Seminar & Exposition, San Antonio, TX, October 19, 2010. http://www.fuelcellseminar.com/media/5499/satyapal_2010_10_19.pdf

[193] International Partnership for Hydrogen and Fuel Cells in the Economy (IPHE), "2010 Hydrogen and Fuel Cell Global Commercialization & Development Update," prepared with support from the U.S. Department of Energy, 2010, http://www1.eere.energy.gov/hydrogenandfuelcells/pdfs/iphe_commercialization2010.pdf; State of California, "California Hydrogen Highway," web page, 2011, http://www.hydrogenhighway.ca.gov/; Honolulu Clean Cities, "Hawaii Hydrogen Initiative," web page, 2011, http://honolulucleancities.org/hawaii-hydrogen-initiative/.

[194] California Fuel Cell Partnership, "Progress," web page, March 2011, http://cafcp.org/progress. A current and interactive map of alternative fuel stations is maintained by the National Renewable Energy Laboratory (NREL) through TransAtlas, available online at http://maps.nrel.gov/transatlas.

[195] Most of these fuel cells are powered by fossil fuels (e.g., natural gas), as opposed to hydrogen. See "Fuel Cell Today: Industry Review 2010," *Fuel Cell Today*, 2010. http://www.fuelcelltoday.com/analysis/industry-review

[196] S. Curtin and J. Gangi, "The Business Case for Fuel Cells," Fuel Cells 2000, Washington, DC, September 2010. http://www.fuelcells.org/info/BusinessCaseforFuelCells.pdf

[197] The project includes 250 kg of storage, approximately 90 kW of electrolysis capacity, a 60 kW hydrogen-fueled internal combustion generator, a 5 kW PEM fuel cell and a hydrogen refueling station for vehicles. NREL, "Wind-to-Hydrogen Project," web page, 2011. http://www.nrel.gov/hydrogen/proj_wind_hydrogen.html

[198] K. Harrison, et al., "Hydrogen Production: Fundamentals and Case Study Summaries," National Renewable Energy Laboratory, NREL/CP-550-47302, presented at the 18th World Hydrogen Energy Conference, Essen, Germany, May 16-21, 2010. http://www.nrel.gov/hydrogen/pdfs/47302.pdf

[199] These are fossil-fuel powered polymer electrolyte membrane (PEM) units. I. Staffell, and R.J. Green, "Estimating Future Prices for Stationary Fuel Cells with Empirically Derived Experience Curves," *International Journal of Hydrogen Energy*, No. 34, 2009, pp. 5617-5628.

In the longer term, the cost reductions anticipated in producing, storing and delivering hydrogen on a large scale for automotive and other markets will also reduce the costs of hydrogen for grid applications. While potentially limited by round-trip efficiency, the economic viability of hydrogen for grid applications will therefore improve as niche fuel cell markets continue to expand in the near term, bringing down costs by achieving economies of scale, mass production and learning-by-doing across the hydrogen infrastructure supply chain.[200]

Technology

Description

Hydrogen can be derived from a variety of sources, using a variety of techniques. **Figure 15** illustrates major hydrogen energy pathways in a simplified schematic. Hydrogen production from electricity by way of electrolysis, is technically feasible. However, lifecycle costs for electrolytic hydrogen tend to be high compared to most of the other hydrogen pathways shown. These pathways have been described in detail elsewhere, with the most common and inexpensive method of producing hydrogen today being steam methane reforming of natural gas, performed at large scale in petroleum refining and fertilizer production industries.[201] Similarly, most grid fuel cell applications today are fueled with natural gas or biogas directly, or with hydrogen derived from natural gas or biogas. Electrolysis units are a commercial technology for numerous niche markets, typically where high purity hydrogen is needed on site and on demand, such as in metallurgy, electronics and generator cooling in large electricity generating power plants. The three main components of an electrolytic hydrogen system, electrolysis, storage, and fuel cells, are discussed in turn below.

---

[200] D. Greene and S. Das, *Bootstrapping a Sustainable North American PEM Fuel Cell Industry: Could a Federal Acquisition Program Make a Difference?* Oak Ridge National Laboratory, ORNL/TM-2008/183, October, 2008.

[201] N. Brinkman et al., *Well-to-Wheels Analysis of Advanced Fuel/Vehicle Systems: A North American Study of Energy Use, Greenhouse Gas Emissions, and Criteria Pollutant Emissions*, Argonne National Laboratory, May 2005; M. Ruth, M. Laffen and T. Timbario, *Hydrogen Pathways: Cost, Well-to-Wheels Energy Use, and Emissions for the Current Technology Status of Seven Hydrogen Production, Delivery, and Distribution Scenarios*, National Renewable Energy Laboratory, NREL/TP-6A1-46612, September, 2009.

Figure 15. Electrolytic and Other Major Hydrogen Energy Pathways

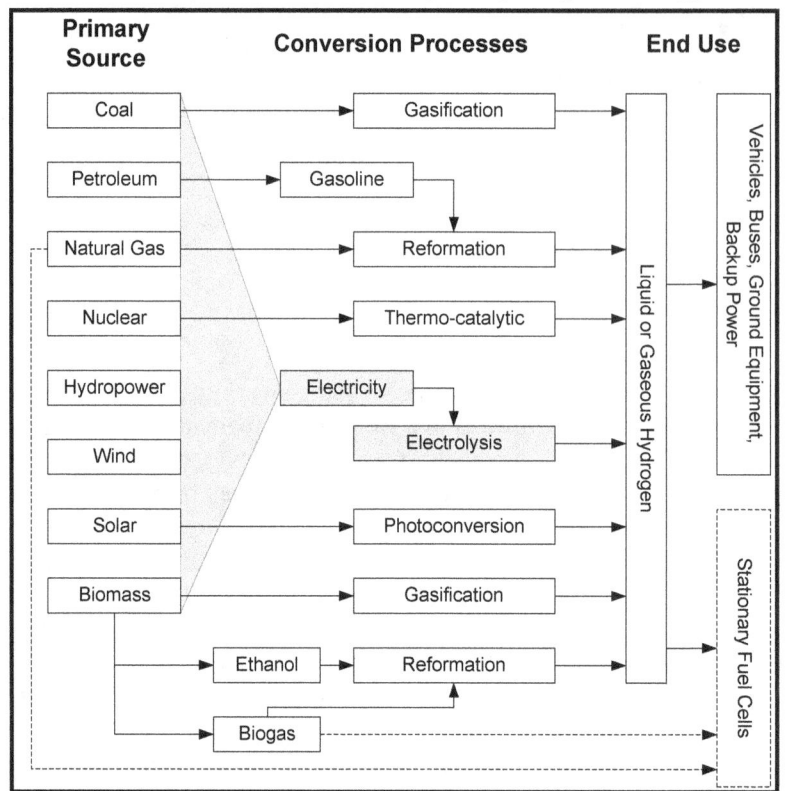

Source: M. Melaina, National Renewable Energy Laboratory, 2011.

Electrolysis

There are two main types of low temperature electrolyzers—alkaline and polymer electrolyte membrane (PEM)—both of which generally operate in the range of 60-80° C.[202] PEM electrolyzers are a fundamentally similar technology to the PEM fuel cells used in automotive applications, but are run in reverse by splitting water molecules with electricity to produce hydrogen and oxygen, rather than combining hydrogen with oxygen to produce electricity and water. Low temperature alkaline and PEM systems are able to meet varying load or demand, and ramp up and down in power levels very quickly (providing multiple grid services).

In a PEM electrolyzer, the hydrogen side (cathode) and the oxygen side (anode) are separated by a solid membrane electrolyte, selectively permeable to hydrogen ions which are transported across the membrane accompanied by water molecules. **Figure 16** illustrates a PEM process. Water reacts at the catalytic surface of the anode to form oxygen, positively charged hydrogen ions (protons), and electrons. Oxygen is collected or released to the atmosphere, the electrons flow through an external circuit, and the hydrogen ions selectively move across the membrane to the cathode. At the cathode, the positively charged hydrogen ions combine with electrons from the external circuit to form hydrogen gas.

---

[202] Polymer electrolyte membrane fuel cells are also called proton exchange membrane fuel cells.

Figure 16. Schematic Representation of PEM Electrolysis

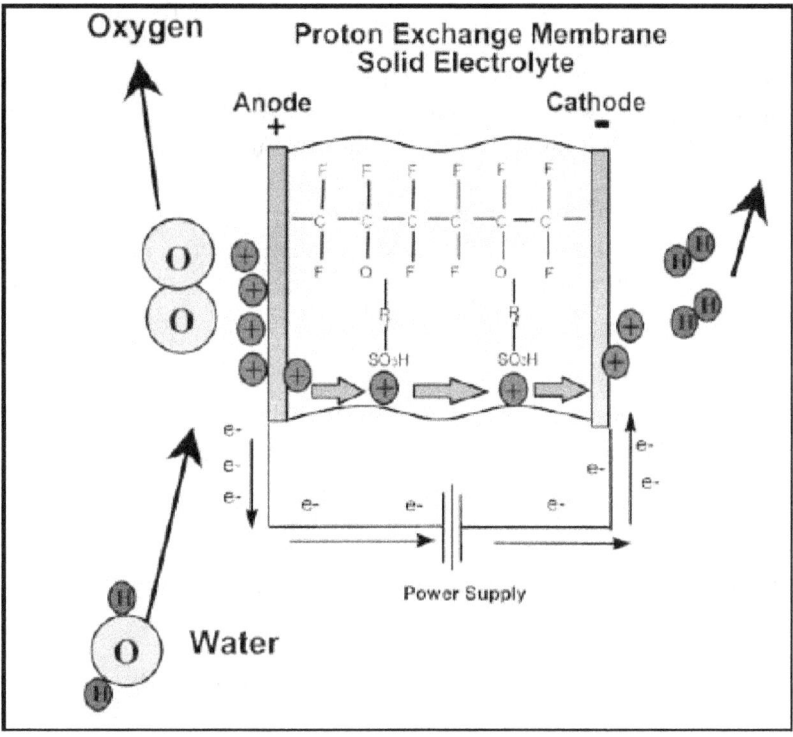

Source: F. Barbir, "PEM Electrolysis for Production of Hydrogen from Renewable Energy Sources," *Solar Energy*, No. 78, 2005, pp. 661-669.

Alkaline electrolyzer systems are the most established electrolysis technology, with several large-scale installations, including one with a reported capacity of 74,000 kg H_2/day.[203] Alkaline electrolyzers use an alkaline solution (potassium hydroxide) as the electrolyte. However, whereas PEM electrolytes transfer hydrogen ions (H+) through a semi-solid polymer membrane, alkaline electrolysis transfers hydroxide ions (OH-) through a hot liquid electrolyte. The PEM or alkaline cell stacks are the core technology of an electrolysis system. Other components include a water supply and circulation system, water-gas separators for hydrogen and oxygen, power supply and voltage regulator, heat exchangers, hydrogen gas drying, and controls.

Hydrogen Storage

There are three general types of hydrogen storage technology: (1) physical storage, including high pressure gas tanks and liquid tanks; (2) geologic storage; and (3) material-based storage, including various types of hydrogen carrier materials.[204] The low volumetric energy density of hydrogen requires very high pressures to store hydrogen in sufficient quantities to provide vehicle ranges comparable to a conventional vehicles.[205] All advanced FCEV designs include physical

---

[203] National Renewable Energy Laboratory, "Current (2009) State-of-the-Art Hydrogen Production Cost Estimate Using Water Electrolysis: Independent Review," NREL/BK-6A1-46676, September 2009.

[204] U.S. Department of Energy, "Hydrogen Storage," fact sheet, October 2006. http://www.hydrogen.energy.gov/pdfs/doe_h2_storage.pdf

[205] Hydrogen has one of the highest energy contents by weight (more than twice that of natural gas), but one of the lowest by volume 10.8 MJ/nm^3 compared to 35.9 MJ/nm^3 for natural gas. U.S. Department of Energy, "Permitting Hydrogen Motor Fuel Dispensing Facilities," PNNL-14518, January 12, 2004, http://www.pnl.gov/fuelcells/docs/ (continued...)

storage, with either compressed hydrogen tanks (at 5,000 or 10,000 pounds per square inch) or liquid hydrogen storage. Automakers have demonstrated onboard hydrogen storage in vehicles with ranges that have approached or exceeded the ~300 mile range typically quoted as desirable by consumers. For example, a Toyota Highlander FCEV demonstrated a 431 mile range under real-world driving conditions, with an average fuel economy of 68.3 miles per gallon gasoline equivalent (mpgge).[206] Other FCEVs have had trouble reaching a 300 mile range. In addition to improving the commercial viability of FCEVs, sufficient onboard storage (and therefore range) enhances the vehicle-to-grid capability of future FCEVs. As an example, a future FCEV with a 100 kW fuel cell system (most likely hybridized with a small battery), and a range of 350 miles, could provide 3 kW of electricity—enough for a large home—for more than 30 hours when starting with a full tank.[207]

Above-ground tanks with lower pressure gas or liquid can be used to store hydrogen for grid applications. Geologic hydrogen storage is technically feasible, as demonstrated by the commercial system maintained to support a 310 mile hydrogen pipeline serving refineries along the Gulf Coast.[208] The three geologic hydrogen storage systems operating today are all in solution-mined salt caverns.[209] Current research activities are attempting to identify additional opportunities in depleted natural gas reservoirs and aquifers.[210] Some of the same formations considered for compressed air energy storage are also being investigated for bulk hydrogen storage.

Material-based storage methods under development include metal hydrides, chemical hydrides, high surface area sorbent materials (including carbon structures), and various types of chemical storage. Laboratory and cost analyses are ongoing to identify the technical and cost potential of numerous types of material-based hydrogen storage.[211]

Fuel Cells

Stored hydrogen can be converted into electricity by means of a combustion engine or fuel cell. The primary focus of government and industry R&D is fuel cells rather than hydrogen combustion engines. Fuel cells combine hydrogen and oxygen to produce electricity, heat, and water through a chemical reaction. Hydrogen is today generally derived from hydrocarbon fuels, such as methane, through a "reforming" process. The grid-oriented fuel cells being actively

---

(...continued)

permit-guides/module2_final.pdf.

[206]H. Lammer, "Eco-Friendly SUV Gets a Hydrogen Mileage Boost," news feature, National Renewable Energy Laboratory, November 13, 2009. http://www.nrel.gov/news/features/feature_detail.cfm?feature_id=1606

[207] Assumes a fuel economy of 65 mpgge, an electrical conversion efficiency of 45%, and a tank capacity of 5.5 kg hydrogen.

[208] Praxair Technology, Inc. "Increase Hydrogen Supply Availability with Cavern Storage," fact sheet, 2006. http://www.praxair.com/praxair.nsf/0/3A0AB529A089B473852571F0006398A3/ $file/027847_PRAX_RefinSpec_4_low_res.pdf

[209] A.S. Lord, "Overview of Geologic Storage of Natural Gas with an Emphasis on Assessing the Feasibility of Storing Hydrogen," SAND2009-5878, Sandia National Laboratory, 2009a.

[210] A.S. Lord, "Investigating the Potential for Hydrogen Geostorage within Igneous and Metamorphic Rocks: A Status Report," Sandia National Laboratory, 2010; A.S. Lord, P. H. Kobos, et al., "A Life Cycle Cost Analysis Framework for Geologic Storage of Hydrogen: A Scenario Analysis," SAND2010-6938, Sandia National Laboratory, October 2010.

[211] U.S. Department of Energy, October 2006.

---

developed employ proton exchange membraness (PEMFC), phosphoric acid (PAFC), molten carbonate (MCFC), solid oxides (SOFC), direct methanol (DMFC), and alkaline (AFC).[212] Some systems run on natural gas or biogas directly (SOFC and MCFC with internal reforming, and PAFC with external, integrated reforming),[213] on hydrogen derived from natural gas or another source (PEMFC, PAFC or AFC), or directly on methanol (DMFC). The main focus for automotive fuel cells is the PEMFC. A primary advantage of PEM fuel cells is their ability to run under variable conditions, including fast-start, on/off cycling, and part-load operation, as well as the ability to operate at low temperatures.[214] Because the focus of PEMFC development is on mobile applications, demonstrated sizes (about 5-100 kW) are much smaller than for non-PEM fuel cells designed for grid applications (about 100-1000 kW).[215]

Performance

The performance of each component within a hydrogen storage system must be evaluated separately. Each component can vary considerably in size and performance depending on the systems configuration and application. Overall, the expected round trip efficiency of hydrogen storage systems is estimated in the range of 28%-41% depending upon technology advances in the near term and the combination of electrolysis, storage, and fuel cell or combustion components.[216]

Electrolysis

Electrolyzer size can range from a few watts to greater than 1 GW depending on design and application. Stacks can be combined in parallel to increase hydrogen production capacity. Alkaline systems are traditionally larger, but PEM manufacturers are trying to scale up design to reach sizes of 1,000 kg H_2/day. Low temperature alkaline and PEM electrolysis systems have conversion efficiencies between 61% and 81% or 65-48 kWh/kg.[217] In the longer term, efficiencies closer to 50-45 kWh/kg could be typical.[218] Compression energy for storage might add about 4 kWh/kg to these values. High temperature electrolyzers, currently under development and based on solid oxide technology, hold the promise of achieving better that 90% electrical efficiency.[219] PEM electrolyzers have the capability of responding rapidly and operating at part

---

[212] Fuel Cell Today, *The Fuel Cell Today Industry Review 2011*, Hertfordshire, UK, September 14, 2011. http://www.fuelcelltoday.com/analysis/industry-review/2011/the-industry-review-2011

[213] For an example of an integrated biogas stationary fuel cell and hydrogen fueling station project, see E. Heydorn,, "Validation of an Integrated Hydrogen Energy Station," Air Products and Chemical, Inc., report to the DOE Fuel Cell Technologies Annual Merit Review, 2010. http://www.hydrogen.energy.gov/pdfs/progress10/viii_5_heydorn.pdf

[214] L. Venturelli, P. Santangelo, and P. Tartarini, "Fuel Cell Systems and Traditional Technologies. Part II: Experimental Study on Dynamic Behavior of PEMFC in Stationary Power Generation," *Applied Thermal Engineering*, Vol. 29, No. 17-18, 2009, pp. 3469-3475.

[215] F. Barbir, and S. Yazici, "Status and Development of PEM Fuel Cell Technology, *International Journal of Energy Research*, Vol. 32, No. 5, 2008, pp. 369-378; Ballard Power Systems, "CLEARgen," specification sheet, April 2011, http://www.ballard.com/files/PDF/Distributed_Generation/CLEARgen_Spec_Sheet.pdf.

[216] D. Steward et al., *Lifecycle Cost Analysis of Hydrogen Versus Other Technologies for Electrical Energy Storage*, NREL/TP-560-46719, National Renewable Energy Laboratory, November 2009.

[217] Ibid. These values assume the higher heating value (HHV) of hydrogen fuel, a measure of heat released during combustion. For an example of an electrolyzer see Hydrogenics Corp., "HySTAT-60," specification sheet, 2011. http://www.hydrogenics.com/assets/pdfs/HySTAT-Q%20Leaflet.pdf

[218] D. Steward, et al., 2009.

[219] Based on the higher heating value (HHV) of hydrogen fuel.

load, so they could potentially act as a variable load for grid services, including operating reserves.[220] However, some units may not be optimized for part load operation, and there are other issues associated with variable operation that include greater chance for hydrogen crossover in low load situations and suboptimal temperature operation which may decrease efficiency.[221] Durability over the stack lifetime is a performance metric that still needs to be demonstrated more thoroughly, especially under variable operation.[222]

Hydrogen Storage

State-of-the-art high pressure automotive hydrogen tank systems have exceeded the 2010 DOE automotive storage goal for gravimetric density (1.5 kWh/kg) and have approached the 2010 goal for volumetric density (0.9 kWh/l).[223] Ongoing R&D efforts are focused on additional cost reductions for onboard storage systems. The ability to rely upon geologic storage for hydrogen, as is done with natural gas today, depends upon a variety of factors, including chemical reactions, contamination, mobility of hydrogen through the formation, and embrittlement and weakening of metals (more of an issue for high pressure systems).[224] Mined salt caverns appear to have the most favorable physical and chemical properties for hydrogen storage, as well as relatively low costs; however, these formations are not ubiquitous across the Unites States as illustrated in **Figure 17**. By using underground formations, hydrogen provides the opportunity for extremely long duration storage that is able to shift energy seasonally, which could be valuable in high penetration renewable generation scenarios.[225]

---

[220] For example, the Proton Onsite HOGEN C30 system has an operating range of 0-100% power. Proton Onsite, "HOGEN C Series Hydrogen Generation Systems," specification sheet, 2011. http://www.protononsite.com/pdf/HOGEN_C.pdf

[221] NREL, "Wind-to-Hydrogen Project," web page, December 5, 2011, http://www.nrel.gov/hydrogen/proj_wind_hydrogen.html; K.W. Harrison et al., The Wind-To-Hydrogen Project: Operational Experience, Performance Testing, and Systems Integration, NREL/TP-550-44082, National Renewable Energy Laboratory, March 2009; F. Barbir, "PEM Electrolysis for Production of Hydrogen from Renewable Energy Sources," *Solar Energy*, Vol. 78, No. 5, 2005, pp. 661-669.

[222] NREL is conducting long-term testing of PEM and alkaline stacks to understand stack performance under both varying and constant operation. See, for example, NREL, "Wind- to-Hydrogen Project," December 5, 2011.

[223] U.S. Department of Energy, *Hydrogen, Fuel Cells & Infrastructure Technologies Program Multi-Year Research, Development and Demonstration Plan*, April, 2009, http://www1.eere.energy.gov/hydrogenandfuelcells/mypp/.

[224] A.S. Lord, 2009a.

[225] D. Steward et al., 2009.

Figure 17. Location of Salt Deposits Across the United States

Source: A.S. Lord, 2009a.

Fuel Cells

Fuel cell projects run from a few watts to greater than 1 MW. Recent efforts have improved fuel cell start-reliability, durability, and cost.[226] Current electrical efficiency of a PEM fuel cell system is approximately 47% with the target efficiency of about 58% based upon the lower heating value (33.3 kWh/kg) of hydrogen.[227] The conversion efficiencies of PEM fuel cells are therefore much higher (2 to 2.5 times) than typical combustion engines. The lifetime for PEM fuel cells in grid applications is about 20,000 hrs, with a DOE target of 60,000 hours by 2020, and the target for vehicles is 5,000 hrs with a peak efficiency of 60%.[228]

Cost

Electrolysis

Electrolysis is an established process, but increased future demand for electrolyzer units will result in capital and operating cost reductions due to higher volumes of production and improved

---

[226] M. Inaba, "Durability of Electrocatalysts in Polymer Electrolyte Fuel Cells," *ECS Transactions*, Vol. 25, No. 1, 2009, pp. 573-581; N.E. Takeuchi, et al., "Investigation and Modeling of Carbon Oxidation of Pt/C under Dynamic Potential Condition," *ECS Transactions*, Vol. 25, No. 1, 2009, pp. 1045-1054; G.S. Tasic, et al. "Non-Noble Metal Catalyst for a Future Pt Free PEMFC," *Electrochemistry Communications*, Vol. 11, No. 11, pp. 2097-2100.

[227] D. Steward, et al., 2009.

[228] U.S. Department of Energy, "Distributed/Stationary Fuel Cell Systems," web page, March 8, 2011, http://www1.eere.energy.gov/hydrogenandfuelcells/fuelcells/systems.html; U.S. Department of Energy, April 2009; D. Steward et al., 2009.

designs. One recent estimate places electrolyzer costs at about $830/kW.[229] Estimates of costs when deployed at scale are in the range of $340/kW-$384/kW for low temperature (PEM and alkaline) electrolyzers. The balance of plant, which includes control systems, power electronics, and auxiliary systems, might be anywhere from 34% to 86% of the capital cost, depending on size, application, and design.[230] Stack replacement costs (at seven year intervals) are estimated at about 35% of total purchased costs.[231] Maintenance costs for low temperature electrolyzers are expected to be relatively low—between 1% and 3% of the total installed capital cost per year for the units.[232]

Hydrogen Storage

Estimates of future costs of high volume manufactured carbon fiber tanks for onboard vehicular storage are $13 and $20/kWh for 5,000 and 10,000 psi tanks, respectively. Similar analyses for future liquid automotive tanks suggest that $8/kWh of storage is achievable.[233] Cost estimates for 2500 psi storage for above ground hydrogen tanks for retail fueling stations are near $900 per kg of usable hydrogen based upon U.S. DOE hydrogen analysis delivery component cost models.[234] Costs for storage of hydrogen in salt caverns has been estimated at $5-$24 per kg of usable hydrogen storage.[235] Use of depleted gas reservoirs, if suitable physically and chemically, would tend to have lower costs.

Fuel Cells

As with other components, hydrogen fuel cells systems are currently not deployed at scale, and there is significant uncertainty as to the cost reduction potential. Current costs for stationary fuel cells are estimated at about $3,000/kW for equipment costs (not including installation).[236] A large fraction of this cost is associated with the platinum catalyst, discussed later in this chapter. Estimates of potential costs after additional R&D and higher deployment volumes are in the range of $400-$800/kW. As with electrolyzers, the small number of moving parts for fuel cells reduces ongoing maintenance requirements. The current O&M costs for fuel cells are estimated at about $50/kW-yr, potentially dropping to $20/kW-yr as the technology becomes more established.[237] One of the largest expenses for stationary fuel cells is the cell stack replacement cost, which is estimated to be about 30% of the initial capital cost after 20,000 hrs of use.[238]

---

[229] D. Steward et al., 2009.

[230] National Renewable Energy Laboratory, September 2009.

[231] Ibid.

[232] Ibid.

[233] S. Lasher, "Analyses of Hydrogen Storage Materials and On-Board Systems," presentation to the Department of Energy Annual Merit Review, Washington, DC, TIAX LLC, June 7-11, 2010. http://www.hydrogen.energy.gov/pdfs/review10/st002_lasher_2010_o_web.pdf.

[234] U.S Department of Energy, "DOE H2A Delivery Analysis," web page, Dec 12, 2011, http://www.hydrogen.energy.gov/h2a_delivery.html.

[235] A.S. Lord, P. Kobos, and D. Borns, *A Life Cycle Cost Analysis Framework for Geologic Storage of Hydrogen*, SAND2009-6310, Sandia National Laboratories, 2009b.

[236] D. Steward et al., 2009.

[237] Ibid.

[238] Ibid.

Research and Development

Research and development activities in the United States are being funded and actively pursued, in part, by the U.S. Department of Energy's Fuel Cell Technologies Program. Some of the technology limitations and research directions can be seen in the program's Roadmap on Manufacturing R&D for the Hydrogen Economy,[239] in a study by the National Academy of Engineering,[240] and in an independent panel review of low temperature electrolysis sponsored by the DOE.[241]

Electrolysis

For electrolysis, areas of interest can be characterized into four broad categories: materials, design improvements, manufacturing, and systems integration/testing.

- **Materials.** Catalyst materials are currently a significant expense for both PEM and alkaline electrolyzers. New catalyst coatings that would reduce the requirement for precious metals could reduce costs, while improved membranes with higher ionic flow and lower resistance would improve efficiency.

- **Design Improvements.** Power conditioning systems that are tuned for specific applications and operation at higher pressure would reduce costs.

- **Manufacturing.** Advances needed include simplified designs adapted for mass production, improved stack forming and assembly, low-temperature metal joining methods, improved in-line quality testing, and improved coating and thin film deposition methods. Research is also needed to develop low-cost sensors and other safety equipment and procedures. Some advances in manufacturing for automotive PEM fuel cell systems may translate to savings for PEM electrolysis systems.

- **System integration.** Improved integration of auxiliary equipment, including power conditioning, impurity removal, water management, and drying, would improve the system energy balance and reduce component costs.

Hydrogen Storage

For storage, research is needed to improve the cost and manufacturability of storage for transportation applications.[242] R&D efforts are needed to develop low-cost, high-pressure tanks whether using conventional materials (where better methods for heat-treating containment vessel junctions and walls adapted to high volume manufacture are needed) or using advanced composites where research and engineering is required to develop lightweight tanks manufacturable at scale. These technologies would be useful for grid applications as well.

---

[239] U.S. Department of Energy, "Roadmap on Manufacturing R&D for the Hydrogen Economy," December 2005. http://www.hydrogen.energy.gov/pdfs/roadmap_manufacturing_hydrogen_economy.pdf

[240] National Academy of Engineering, *The Hydrogen Economy: Opportunities, Costs, Barriers, and R&D Needs*, National Academies Press, 2004.

[241] National Renewable Energy Laboratory, September 2009.

[242] U.S. Department of Energy, December 2005.

Investigation of geologic storage may reveal opportunities for lower-cost and larger-scale hydrogen storage systems for utility applications.[243]

Fuel Cells

For fuel cells, many of the R&D needs are similar to those for electrolyzers. Key issues for fuel cells fall into the categories of materials, design, and manufacturing.[244]

- **Materials.** R&D needs include membrane research for operation at low relative humidity and high temperatures with greater durability and lower cost, electrode catalyst loading with either reduced precious metal requirements or alternative materials with lower cost and similar performance, bi-polar plates with improved weight and size, and corrosion resistant coatings for supports, which would lengthen stack life.

- **Design.** Better power electronics subsystems, as well as air, thermal, and water management systems can help reduce cost and increase performance.

- **Manufacturing.** Improved manufacturing provides the greatest opportunity for cost reduction and is key to the deployment of large numbers of fuel cell systems. High speed and automated manufacturing R&D needs include sealing and coating methods for electrodes, membranes, and bi-polar plates as well as high speed component production. Manufacturing issues are especially important to successful introduction of FCEVs, which would require much greater volumes than most grid-oriented fuel cell applications. Tied to manufacturing are testing needs which reduce the time, footprint, and equipment required for testing, break-in, and acceptance of a system. This includes high speed quality control and defect detection. Low-cost sensors and safety equipment are needed for all stages of production and operation.

Deployment Challenges

Hydrogen faces a number of deployment challenges beyond those associated with cost and efficiency. The most significant is the development of a new fueling infrastructure for hydrogen vehicles. Hydrogen is currently stored and delivered as a process gas for some industrial uses via pipelines and steel tank tubes, but the necessary infrastructure for widespread use, especially in transportation, would require a many-fold expansion of this infrastructure. Small quantities of hydrogen can be transported by truck, but this method quickly becomes cost-prohibitive at high volumes. Although hydrogen pipelines exist, they are not widely distributed. Development costs for an extensive pipeline infrastructure would be high, especially during early market introduction when demand for hydrogen is low.[245] Hydrogen can be blended with natural gas at low volumes (less than 10%-20%) with little concern for safety or appliance compatibility, and some newer

---

[243] A.S. Lord, 2009a.

[244] U.S. Department of Energy, April 2009; J. Nie, and Y. Chen, "Numerical Modeling of Three-Dimensional Two-Phase Gas-Liquid Flow in the Flow Field Plate of a PEM Electrolysis Cell," *International Journal of Hydrogen Energy*, Vol. 35, No. 8, 2010, pp. 3183-3197; Tasic et al., 2009.

[245] Hydrogen pipelines differ from natural gas pipelines since hydrogen can cause steel embrittlement and may require different seals and fittings. J.L. Gillette and R.L Kolpa, *Overview of Interstate Hydrogen Pipeline Systems*, ANL/EVS/TM/08-2, Argonne National Laboratory, 2007.

---

natural gas pipelines could be converted to hydrogen pipelines if made of the right steels or plastics.[246] Near term developments that do not require extensive hydrogen infrastructure, such as production onsite at distributed refueling stations, can facilitate the adoption of hydrogen as a vehicle fuel and energy carrier. Early market applications are particularly important to achieve manufacturing scale-up and reduce costs.

Safety is a concern, primarily with hydrogen being used as a transportation fuel. Hydrogen's physical and chemical characteristics have both safety benefits and drawbacks in its production, transportation, and use as a fuel.[247] Hydrogen's safety advantages compared to other fuels include extremely fast dispersal after a leak compared to gasoline, higher flammability limit than gasoline (4% compared to 1%), lower risk that secondary materials will ignite, and no toxic gases produced from burning. The primary drawbacks of hydrogen in comparison to other fuels are its relatively high probability of ignition and the combination of low ignition energy and wide flammability range, resulting in a higher risk of ignition from leak points in equipment.[248]

Hydrogen burns with a flame not visible in daylight and has no odor, increasing the concern of undetected leaks leading to hazardous fire conditions. High pressure hydrogen systems are prone to leakage due to the physical properties of hydrogen, such as small molecular size, but these same properties ensure a rapid rate of dispersal noted above—although special precautions must be taken for indoor, contained systems. Explosive conditions can occur in contained systems at concentrations in air of approximately 18%, but these concentrations are nearly impossible to achieve in outdoor systems. Common odorants and flame enhancers poison fuel cells, so specialized sensors are required for detection; research into alternative odorants is ongoing. As is the case with natural gas vehicles and tank trucks, hydrogen storage tanks have a low probability of release in collisions during transportation because of the high impact strength of compressed gas tubes. Overall, different precautions must be taken to ensure hydrogen achieves the same level of safety as other fuels.[249]

Both electrolyzers and fuel cells depend on expensive materials with varying degrees of availability. The electrolyzer cell stack often includes expensive noble metals for the cathode and anode where the hydrogen and oxygen are formed. Typically, these include platinum, iridium, and ruthenium based catalysts, which are needed to work within an acidic, corrosive environment. The catalyst loading for PEM electrolyzers might be anywhere from 0.4 mg/cm^2 to a few mg per square centimeter with the cell active area for a 1,000 kg/day unit being greater than 2,000 cm^2.[250]

---

[246] NaturalHy Project, "About NaturalHy," web page, 2011. http://www.naturalhy.net/index.php?option= com_content&view=article&id=45&Itemid=28

[247] S.E. Plotkin, "Assessment of PNGV Fuels Infrastructure: Infrastructure Concerns Related to the Safety of Alternative Fuels," ANL/ESD-TM-160, Argonne National Laboratory, June 2000.

[248] U.S. Department of Energy, "Hydrogen Data," DOE/GO12008-2597, October 2008. http://www.hydrogen.energy.gov/permitting/pdfs/43061.pdf

[249] For additional safety information, see C. Grant, *Reaching the U.S. Fire Service with Hydrogen Safety Information: A Roadmap*, Prepared by the Fire Protection Research Foundation for the National Renewable Energy Laboratory, 2009.

[250] L. Ma, et al., "Investigations on High Performance Proton Exchange Membrane Water Electrolyzer." *International Journal of Hydrogen Energy*, Vol. 34, No. 2, 2009, pp. 678-684; P. Millet, et al., "GenHyPEM: A Research Program on PEM Water Electrolysis Supported by the European Commission, " *International Journal of Hydrogen Energy*, Vol.34 No. 11, 2009, pp. 4974-4982; J. Cheng, et al., "Study of Carbon-Supported IrO2 and RuO2 for Use in the Hydrogen Evolution Reaction in a Solid Polymer Electrolyte Electrolyzer," Electrochimica Acta, Vol. 55, No. 5, 2010, pp. 1855-1861; J.D. Holladay, et al., "An Overview of Hydrogen Production Technologies," *Catalysis Today*, Vol. 139, No. 4, 2009, pp. 244-260; P. Millet, et al., "PEM Water Electrolyzers: From Electrocatalysis to Stack Development," International Journal of Hydrogen Energy, Vol. 35, No. 10, May 2010, pp. 5043-5052.

Alkaline electrolyzers often operate at lower current density which requires even larger cell active areas, 15,000 cm^2 or greater, for 1,000 kg/day hydrogen production. PEM fuel cells typically require platinum based catalysts and can contribute to more than 50% of the cell stack cost.[251] The platinum loading for the PEM fuel cells might be 0.15 mg/cm^2 or greater, or 170-214 mg/kW.[252] The hydrogen production process requires approximately 3 gallons of water per kg H_2 while cooling might require anywhere from 0.1-300 gallons of water per kg H_2 depending on the size of the production plant and the type of cooling system implemented.[253]

Other potential impacts (such as land use) have yet to be quantified in detail, although are unlikely to be significantly larger than other storage technologies.

Conclusions

Hydrogen systems are distinct from other storage systems in the ability to serve both grid and transportation energy markets. However, hydrogen storage and fuel cell technologies face a number of technical and economic challenges before becoming competitive. Electrolytic hydrogen systems are currently both more expensive and have lower round-trip efficiencies than a number of commercially available technologies for grid applications. For transportation, FCEVs face the challenge of cost, durability and lack of a refueling infrastructure, though infrastructure developments are proceeding in California, Hawaii, and internationally. Many key hydrogen technologies, particularly fuel cells, are still in the early stages of commercialization, and will require substantial cost reductions to achieve large-scale and economical deployment. Progress in niche markets, such as forklifts for warehouses and small-scale backup power for telecom sites, is pushing the technology forward and resulting in reduced costs as the volume of units purchased increases.

---

[251] G.S. Tasic, et al., 2009; S. Zhang, et al., "A Review of Platinum-Based Catalyst Layer Degradation in Proton Exchange Membrane Fuel Cells," Journal of Power Sources , Vol. 194, No. 2, 2009, pp. 588-600; S. Lasher, "Direct Hydrogen PEMFC Manufacturing Cost Estimation for Automotive Applications," presentation to the Department of Energy Annual Merit Review, Washington, DC, TIAX LLC, May 17, 2007, http://www.hydrogen.energy.gov/pdfs/ review07/fc_27_lasher.pdf.

[252] S. Harmin, et al., "Synthesis of Novel Electro-catalysts for Proton Exchange Membrane Fuel Cells," *Separation Science & Technology*, No. 38. p. 2963, 2003; S.A. Grigoriev, et al., "On the Possibility of Replacement of Pt by Pd in a Hydrogen Electrode of PEM Fuel Cells," *International Journal of Hydrogen Energy*, Vol. 32, No. 17, 2007, pp. 4438-4442; B. James, et al., "Mass-Production Cost Estimation for Automotive Fuel Cell Systems," *DOE Hydrogen and Fuel Cells Program, FY2011 Annual Progress Report*, 2010, pp. 609- 613.

[253] For instance, the Hogen C30 would consume 7.1 gal/hr of de-ionized water and require liquid cooling of both fluids (up to 35 gal/min) and the hydrogen dryer sub system (up to 20 gal/min); Barbir, 2005; National Renewable Energy Laboratory, September 2009.

Chapter 7: Compressed Air Energy Storage

Overview

Compressed Air Energy Storage (CAES) is a commercially available, utility scale, bulk electricity storage technology that uses high-pressure air as a storage medium. Large-scale, airtight storage volumes can be developed in geologic formations such as underground salt domes and saline aquifers. In conventional "diabatic" CAES, stored compressed air is released through a modified gas turbine, requiring the use of natural gas, making CAES a hybrid storage/generation technology. CAES is typically considered to have the lowest capital cost of any bulk electricity storage technology. The first CAES plant was completed in 1978 in Huntorf, Germany. It was designed primarily to provide "black start" capability (provide a source of power to start conventional generators after a system-wide power failure), and it was rated at 290 MW with two hours of capacity.[254] A second plant (**Figure 18**) was built in 1991 in McIntosh, AL, and has a rating of 110 MW for 26 hours.[255]

Figure 18. 110 MW CAES Plant in McIntosh, AL

Source: Courtesy of PowerSouth Energy Cooperative, 2006.

CAES can be scaled to several-hundred megawatts or even gigawatts. CAES can provide long duration storage with independently scalable energy storage capacity and can provide efficient operation over a broad range of operating conditions. These characteristics make CAES beneficial on multiple time scales and able to provide many services, including operating reserves and load

---

[254] F. Crotogino, et al., "Huntorf CAES: More Than 20 Years of Successful Operation," Solution Mining Research Institute Spring 2001 Meeting. Orlando, FL, March 22, 2001.

[255] R. Schalge, and B. Mehta, "The Alabama Electric Compressed Air Storage Cavern from Planning to Completion," Proceedings of the American Power Conference, Chicago, IL, 1993.

following, generally mitigating the impact of large-scale ramp events from variable generation resources.

Conventional diabatic CAES is subject to a set of detailed siting criteria associated with its use of underground geological formations. Without a comprehensive study of the availability of geology suitable for CAES it would be difficult to assess the extent to which project siting could limit CAES deployment. Several specific CAES projects are in various stages of development and developers have indicated interest in exploring the possibility of developing additional ones. The success of future projects depends on the ability of utilities, developers, and regulators to address existing deployment barriers. **Table 7** provides a list of several proposed CAES plants in the United States This list is not comprehensive, as there have been at least preliminary analysis and proposals for other CAES facilities in Texas, Montana, Utah, North Dakota, and Arizona.

Table 7. Proposed CAES Plants in the United States

Name	Location	Cavern Type	Capacity (MW)
McIntosh Power Plant (existing)	McIntosh, Alabama	Salt dome	110
Iowa Stored Energy Park[a]	Dallas Center, Iowa	Aquifer	135-270
Norton Energy Storage[b]	Norton, Ohio	Abandoned hard rock mine	2700
PG&E[c]	Kern County, California	Porous rock	300
Seneca (NYSEG/Iberdrola)[d]	Schuyler County, NY	Bedded salt	150

Source: NREL compilation from:

a. K. Holst and M .King, "Iowa Stored Energy Park," presentation to the U.S. Department of Energy, Energy Storage Systems Program Update Conference 2010, Washington, DC, November 2, 2010; Iowa Stored Energy Park, "Frequently Asked Questions," web page, Dec. 14, 2011, http://www.isepa.com/FAQs.asp.

b. Norton Energy Storage L.L.C., "Application to the Ohio Power Siting Board for a Certificate of Environmental Compatibility and Public Need," Summit County, OH, 2000.

c. H. LaFlash "Compressed Air Energy Storage" slide presentation, Pacific Gas and Electric Company, Nov 3, 2010, http://www.sandia.gov/ess/docs/pr_conferences/2010/laflash_pge.pdf

d. J. Rettberg, "Seneca Advanced Compressed Air Energy Storage (CAES) 150MW Plant Using an Existing Salt Cavern," slide presentation, November 3, 2010, http://www.sandia.gov/ess/docs/pr_conferences/2010/rettberg_nyseg.pdfNYSEG.

Technology

Description

The operation of a CAES system mirrors that of a combustion turbine except that compression and expansion are decoupled in time. Conventional CAES technology (illustrated in **Figure 19**) uses grid electricity to run a compressor "train" that raises air to high pressure through several stages and injects the air into storage, typically an underground geologic formation. The two existing CAES facilities use salt domes, where the underground cavity was formed by solution mining—pumping fresh water into the formation to dissolve the salt, and pumping out the resulting brine for disposal or other use.[256] Other formations have been proposed, discussed later

---

[256] R.L. Thoms, and R.M. Gehle, "A Brief History of Salt Cavern Use," AGM, Inc., 2000.
(continued...)

in this chapter. Intercooling is typically used at the compression stage to bring the high-pressure air back to near-ambient temperatures prior to injection to reduce storage volume requirements and minimize thermal stress on the geologic formation. Electric power can be regenerated by withdrawing air from storage, combusting fuel (typically natural gas) and expanding the combustion products through a turbine (often in two stages).

While conventional diabatic CAES has been the only technology deployed thus far, several other variants have been proposed. These are often focused on alternative methods for managing the heat of compression and schemes for reducing the fuel consumption. By storing the heat generated during compression to reheat the air at a later time or by compressing the air isothermally, the need to intercool the compressor train and to combust fuel to heat air withdrawn from storage can be reduced or eliminated.[257] In addition, better heat integration and novel turboexpander design can help eliminate the need for a specialized combustor on the high-pressure turboexpansion stage.[258] Additionally, storage of air in aboveground vessels for small-scale applications has been investigated as a way to circumvent the subsurface engineering requirements of siting a CAES facility (especially for small-scale applications).[259]

Figure 19. CAES System Diagram

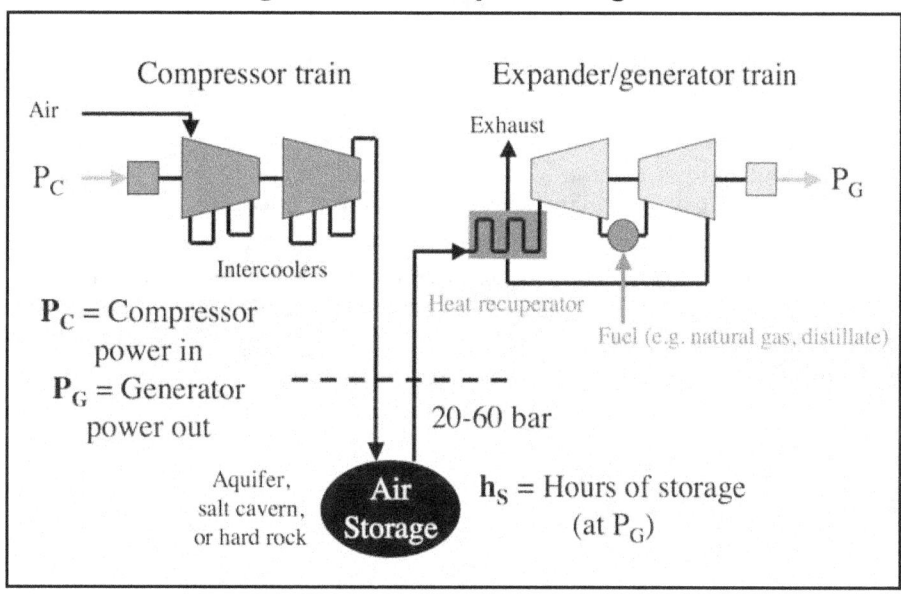

Source: S. Succar and R.H. Williams, *Compressed Air Energy Storage: Theory, Operation and Applications*, Princeton Environmental Institute, Princeton University, 2008.

(...continued)

http://www.solutionmining.org/assets/files/BriefHistory.pdf

[257] C. Bullough, et al., "Advanced Adiabatic Compressed Air Energy Storage for the Integration of Wind Energy," Proceedings of the European Wind Energy Conference, EWEC 2004, European Wind Energy Association, London, UK, November 22-25, 2004.

[258] I. Tuschy, et al., "Evolution of Gas Turbines for Compressed Air Energy Storage," *VGB Powertech*, Vol. 85, No. 4, 2004, pp. 84-7.

[259] EPRI/DOE, 2003.

Performance

CAES is a utility-scale technology, with its size determined by the availability of turbo-machinery equipment and reservoir availability. CAES plants consist of one or more "blocks" of expander capacity, each in the range of 100-300 MW. Having multiple blocks increases the flexibility of the plant since each can be run independently. Because diabatic CAES output is managed by regulating air flow rather than inlet temperature, as in a conventional combustion turbine, CAES has comparatively high part load efficiencies and ramp rates, potentially enabling a wide range of grid services such as voltage support, ramping, and frequency regulation. The existing U.S. plant has a single turbomachinery drive train using a common motor-generator set connected to the compressor and expander via clutches. This results in turnaround times from compression to expansion of approximately 30 minutes, limiting its use in providing operating reserves and other services requiring fast response.

Proposed CAES plants include a dedicated motor drive compressor and expander-generator that would eliminate the single turbomachinery train.[260] This would allow for faster switchover from compression to generation, thus increasing its usefulness for providing ancillary services and responding to increased variability of net load. Once operating, CAES plants can provide rapid ramp rates; the McIntosh plant is capable of ramping at about 18 MW (16% of full output) per minute, or rates that are more than 50% greater than a typical gas turbine.[261] While CAES can provide a variety of services, including those needed to aid in renewables integration, it is also well suited to help address transmission constraints in locations with good wind resources. CAES has been proposed to reduce curtailment due to transmission constraints by co-locating it with wind generation, allowing for greater utilization of transmission lines.[262] As stated earlier in this report, this application could decrease the amount of new transmission needed to access remote wind resources, but represents a trade-off between the most cost-effective use of storage, and the cost of new transmission. In general, it is not optimal to co-locate wind turbines and storage because doing so decreases the usefulness of the storage device. However, in cases where transmission is difficult (or impossible) to site, or is very expensive, the loss of opportunities for storage may be exceeded by its use as an alternative to transmission.[263]

---

[260] Norton Energy Storage L.L.C., 2000.

[261] S. Succar and R.H. Williams, 2008.

[262] N. Desai et al., "Study Of Electric Transmission In Conjunction With Energy Storage Technology," Lower Colorado River Authority, Texas State Energy Conservation Office, August 21, 2003.

[263] P. Denholm and R. Sioshansi, "The Value of Compressed Air Energy Storage with Wind in Transmission-Constrained Electric Power Systems," *Energy Policy*, No. 37, pp. 3149-3158.

Since CAES is a hybrid system, its efficiency cannot be simply stated as a single number.[264] The performance of CAES is based on two separate quantities: the heat rate and the charging electricity ratio. The heat rate is the fuel consumption per unit electrical output similar to heat rates quoted for conventional thermal generation. A typical value for the diabatic CAES heat rate is approximately 3900-4400 Btu/kWh, although improved turboexpansion cycles and heat recovery can reduce fuel consumption further.[265] The charging electricity ratio (CER) is the ratio of output electricity to input electricity of the plant, typically in the range of 1.2-1.8, with higher values for increased pressure ratios across the turboexpander train and greater numbers of compression stages. The fact that the charging ratio is greater than one means that CAES produces more electricity than it consumes, with the balance made up by the consumption of natural gas. As a result, the production of 1 kWh from a CAES plant requires the input of about 0.6 to 0.8 kWh of electricity and 3900-4400 Btu of natural gas fuel. The CER also takes into account piping and throttling losses (a function the reservoir pressure range) as well as compressor and expander efficiencies. Turbine efficiency is especially important in the low-pressure expansion stage where approximately three quarters of the power is generated.[266] Increased turbine inlet temperatures (e.g., by using expander blade cooling technologies) would enhance the turbine and CAES electrical efficiencies.[267] Unlike almost all other storage technologies, there is virtually no decay or self-discharge of stored energy, at least when deployed in salt domes, since these formations are self-healing, meaning pores on the cavity walls seal themselves with available air moisture, virtually eliminating the possibility of air leakage.

Figure 20 provides a method to compare the impact of the heat rate and CER of CAES with conventional storage plant efficiency (such as in a battery or pumped hydro plant). It provides the dispatch price, or fuel related costs (both natural gas and electricity) for generation of electricity from both a CAES plant and a conventional storage plant with different natural gas prices and efficiencies. Ignoring O&M costs, the dispatch price of CAES is the sum of the cost of electricity for compression, plus natural gas, while for a conventional storage plant it is just the cost of charging electricity, considering storage losses. At very low off-peak electricity prices, conventional storage has an advantage over CAES due to the fixed natural gas cost, while at higher off-peak prices, CAES has an advantage due to the need for less electricity purchases. It should be noted that this chart does not consider the capital cost of the device, where CAES has additional advantages over many other storage technologies.

---

[264] Expressing the efficiency of CAES as a single number is more of an academic exercise than useful for estimating its economic performance. There are many ways of calculating a single efficiency number with a large range of values. Since a single value for efficiency is often "insisted on," the CAES community typically expresses the round trip efficiency of CAES in the range of 70%-85%. This value is based on assigning an electrical equivalency to the natural gas input, assuming it would have otherwise be used in a natural gas turbine with an efficiency in the range of 30%-40%. Lower net efficiency values, often below 50% are cited by those who combine the input energy of natural gas and electricity equally, ignoring the relative quality of energy inputs. Because the thermal energy in the fuel input is subject to some conversion efficiency that is independent of the roundtrip storage efficiency of CAES, it is not appropriate to sum the electricity and fuel inputs in the denominator when calculating the plant efficiency. This issue is discussed at length in S. Succar and R.H. Williams, *Compressed Air Energy Storage: Theory, Operation and Applications,* Princeton Environmental Institute, 2008. Detailed thermodynamic analysis of CAES plants is provided by P. Zaugg, "Energy Flow Diagrams For Diabatic Air-Storage Plants," Brown Boveri Review, Vol. 72, No. 4, 1985, pp. 179-183.

[265] Btu=British thermal units.

[266] D.R. Hounslow, et al., "The Development of a Ccombustion System for a 110 MW CAES Plant," Journal of Engineering for Gas Turbines and Power-Transactions of the ASME, Vol. 120, No. 4, 1998, pp. 875-883.

[267] I. Tuschy, et al., "Compressed Air Energy Storage with High Efficiency and Power Output," *VDI Berichte,* No. 1734, 2002, pp. 57-66.

Figure 20. Comparison of CAES Dispatch Cost to Conventional Storage

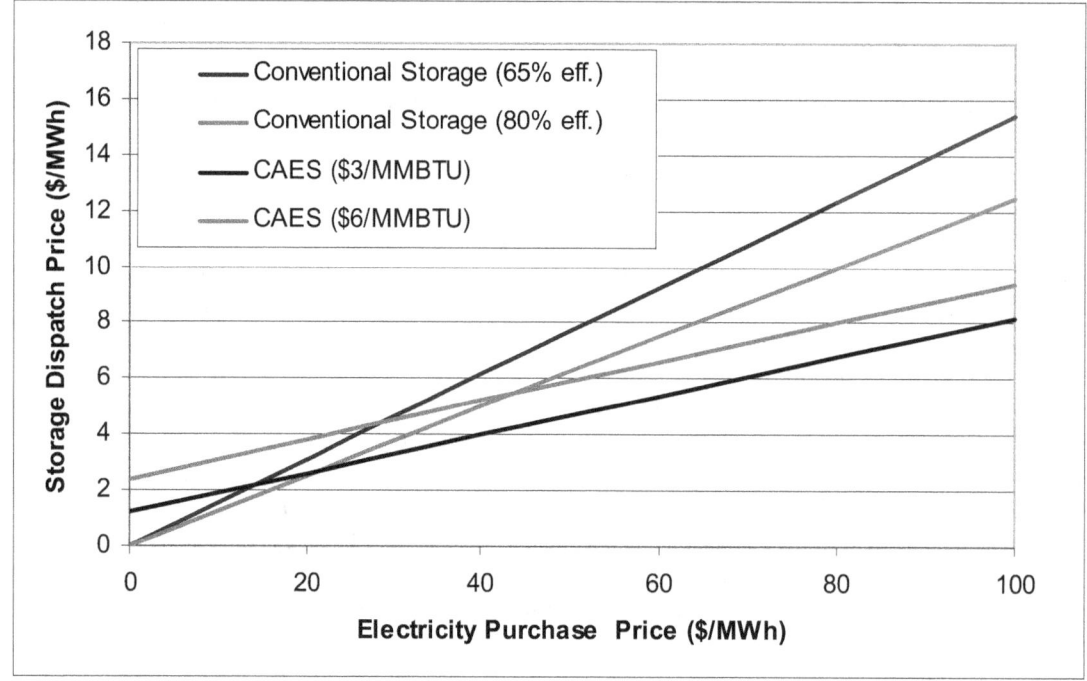

Source: P. Denholm, National Renewable Energy Laboratory.[268]

The lifetime of CAES is similar to that of a combustion turbine. While operation and maintenance is required to replace turbine blades, well materials, and other components, the existing systems at Huntorf and McIntosh continue to operate after 20 to 30 years with very high running and starting availabilities (over 95%).[269] The impact of rapid ramping and multiple daily starts might increase operation and maintenance costs, but the conventional CAES is a proven technology with a well-established operational record.

Cost

Given that no CAES facilities have been constructed since 1991, it is difficult to accurately estimate the current capital cost of the technology. Recent estimates for the capital cost of conventional diabatic CAES (**Figure 21**) range from $600 to $1200/kW with a mean range of $880 to $1020/kW.[270]

---

[268] The dispatch cost of conventional storage (pumped hydro, batteries etc.) is the electricity purchase price divided by the efficiency. The dispatch cost of CAES is the electricity purchase price multiplied by the energy ratio plus the cost of natural gas multiplied by the heat rate. In this figure the energy ratio of CAES is assumed to be 0.7 and a heat rate of 4000 BTU/kWh.

[269] G. Lucas and H. Miller, "Dresser-Rand SmartCAES Technology," Integrating Wind-Solar-CAES, 2nd Compressed Air Energy Storage (CAES) Conference & Workshop, Columbia University, New York, NY, October 20, 2010.

[270] The cost estimates provided are for large-scale CAES systems (100-300 MW). Smaller, distributed CAES systems (10-20 MW) using aboveground storage vessels will be a factor of 2 to 3 higher in total overnight capital cost and roughly two orders of magnitude higher in terms of the capital cost for incremental storage capacity.

Figure 21. Capital Cost Estimates for Conventional Diabatic CAES

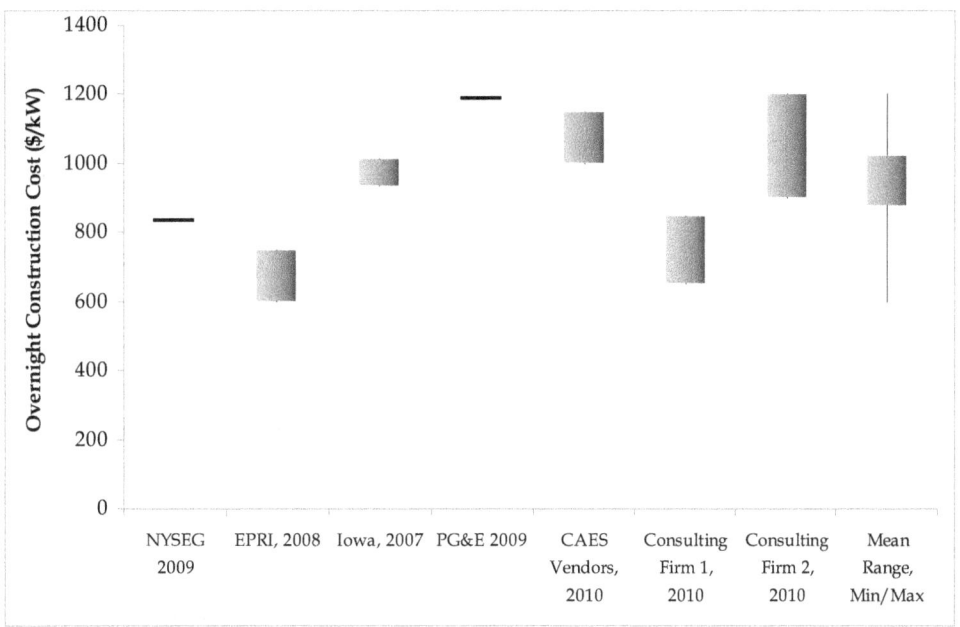

Source: S. Succar, Natural Resources Defense Council. Compiled from references in **Table 7** and other published and unpublished sources.

Total capital cost for CAES can be disaggregated into surface turbomachinery costs and subsurface costs with the latter projected to account for approximately 5%-10% of total project capital cost depending on the type of geology chosen.

**Table 8. Component Costs of a Conventional CAES System
Deployed in a Salt Cavern**

Component	Cost ($/kW)	Fraction of Total
Compressor	87	11%
Heat exchanger	34	4%
High pressure expander	62	8%
Low pressure expander	144	19%
Electrical	45	6%
Construction, labor, indirect costs	324	42%
Cavern development	77	10%
Total	774	100%

Source: R.B. Schainker, and A. Rao, *Compressed Air Energy Storage (CAES) Scoping Study for California*, CEC-500-2008-069, Electric Power Research Institute for California Energy Commission, 2008.

To gauge the potential for future cost reduction associated with CAES turbomachinery, it is common to use the natural gas combustion turbine (CT) as a reference or lower bound. CT capital costs of $650-$690/kW imply a CAES-equivalent cost lower bound of $220-$230/kW.[271] While it is unlikely that CAES surface turbomachinery will reach these costs in the near term, it does suggest high levels of deployment could drive substantial capital costs reductions.

Typical numbers for variable and fixed O&M for CAES are $60/kW-year and $13/kW-year respectively.[272] Low duty cycle operation with high frequency switching between compression and expansion modes of operation could considerably increase the variable O&M and the levelized replacement costs for critical components of the turboexpander train.

Research and Development

Conventional CAES is considered a fairly mature technology, based on well proven gas-turbine technology and with two plants operating for over two decades. There are four general areas of RD&D activities, listed below, with the first two being near-term activities, and the second two being more research oriented.

More Optimized CAES Equipment

Previous CAES plants used components that were not optimized for the unique characteristics of the CAES expansion cycle. This is partially due to the small market for which developing dedicated equipment would not be worthwhile. A large CAES market could drive development of custom turbo-machinery, improving the efficiency of CAES components. While the existing plants in Huntorf and McIntosh are based largely on legacy steam and gas turbine technology, it is likely that future system designs will benefit from improved heat integration, more specialized surface turbomachinery, better systems integration and newer cycles that take advantage of the technology's unique operating characteristics. CAES output is changed by adjusting the air flow (taking advantage of the compressed air from storage) rather than by adjusting the inlet temperature as in a conventional gas turbine. This contributes to the overall low heat rate and high part load efficiency of CAES, but also allows for more specialized cycles to be employed to take advantage of these characteristics.[273] One such example of a new design could provide improved heat recovery and elimination of specialized combustors.[274]

Near-term cost reductions can be realized by using CAES designs that rely to a greater extent on off-the-shelf technology.[275] Several variations of this approach have been explored recently as

---

[271] CT costs are from U.S. Energy Information Administration, *Annual Energy Outlook 2010*, DOE/EIA-0383(2010), 2010. Because CTs divert 2/3 of their turboexpander output to power the compressor, construction cost expressed as dollars per net output capacity reflects only one third of the CT turbine's capacity. By comparison, because power from the grid powers the CAES compressor, the full output of the turboexpander is reflected in its construction cost. This means that a factor of 1/3 should be applied to the cost of the CT in order to make an equivalent comparison with CAES capital costs

[272] EPRI/DOE, 2003.

[273] It should be noted that the heat rate of a CAES plant is much lower than a simple-cycle CT combustion turbines (8,600-10,700 Btu/kWh), and has better part load efficiency. It is more appropriate to compare a CAES plant to a CT, since it is more likely to provide load-following, peaking, and ancillary services. While combined cycle systems have much lower heat rates (6,300-7,200 Btu/kWh), they are somewhat less suited to perform these services.

[274] Tuschy, I., et al., Compressed air energy storage with high efficiency and power output. VDI Berichte, 2002(1734): p. 57-66.

[275] Nakhamkin, M., "Second Generation of the CAES Technology," Center for Life Cycle Analysis Compressed Air Energy Storage Scoping Workshop, Columbia University, October 21, 2008.

---

exemplified by plans for the 150 MW CAES plant proposed by the New York State Electric & Gas Corporation.[276] By integrating a stand-alone CT into the CAES configuration, capital costs can be significantly reduced (estimated by the company to be roughly half the cost of a conventional diabatic CAES system). However, these near-term cost improvements might be offset somewhat by a higher heat rate and/or reduced CER, diminished operational flexibility or reduced potential for further cost reduction as discussed above.

Demonstrate CAES in New Underground Formations

The only deployment of CAES to date has been in salt domes, which are only available in a small fraction of the United States.[277] A frequently cited barrier to CAES deployment is the perceived difficulty of finding geology suitable, but available data is insufficient to accurately determine whether this is the case. Although CAES can theoretically use a wide variety of geologic formations, current utility-scale experience is limited. Demonstration of CAES at commercial scale over a broad range of geologies could significantly accelerate deployment in the near term. The development of commercial CAES projects in saline aquifers, depleted gas wells, abandoned mines, compensated hard rock caverns and salt beds would reduce perceived project development risk of CAES and aid its adoption in the marketplace. Utilities to date have been unwilling to develop CAES under this uncertainty. Projects now being supported by ARRA and state agencies will demonstrate CAES in the major geology types, including bedded salt in New York, an aquifer in Iowa, and a depleted gas field in California (**Table 7**).

Reduce or Eliminate Fuel Use

There are two pathways being pursued to reduce use of fuel in CAES systems. The first is adiabatic CAES, where the heat of compression is stored for later use in the expansion cycle.[278] Relatively little work has been performed on adiabatic CAES in the United States beyond engineering studies. This approach is currently being pursued primarily in Europe, with at least one proposed plant,[279] driven in part by concerns about CO_2 emissions and natural gas supply security concerns, especially since adiabatic cycles are unlikely to be cost effective without high natural gas prices.[280] A CAES plant that does not burn natural gas can use efficiency as the primary performance metric (similar to other bulk energy storage technologies), with estimates of

---

[276] New York State Energy Research and Development Authority, "Compressed Air Energy Storage Engineering and Economic Study," Final Report 10-09, prepared by New York State Electric and Gas, December 2009; J. Rettberg, 2010.

[277] Domal salt is limited to a few gulf coast states. Bedded salt is available in a few more states, but the dominant source of underground caverns for future deployment would need to be porous rock. See S. Succar and R.H. Williams 2008 for a map of the United States showing the areas of suitability for potential CAES development.

[278] G. Grazzini and A. Milazzo, "Thermodynamic Analysis of CAES/TES Systems for Renewable Energy Plants," *Renewable Energy*, No. 33, 2008, pp. 1998–2006; E. Macchi and G. Lozza, "A Study of Thermodynamic Performance of CAES Plants, Including Unsteady Effects," TP 87-GT-23, International Gas Turbine Conference and Exhibition., Anaheim, CA, May 31, American Society of Mechanical Engineers, 1987.

[279] RWE Power AG, "ADELE—Adiabatic Compressed-Air Energy Storage for Electricity Supply," January 2010. http://www.rwe.com/web/cms/mediablob/en/391748/data/364260/1/rwe-power-ag/innovations/adele/Brochure-ADELE.pdf

[280] R.W. Reilly and D.R. Brown, "Comparative Economic Analysis of Several CAES Design Studies," Proceedings of the Intersociety Energy Conversion Engineering Conference, Atlanta, GA, American Society of Mechanical Engineers, January 1, 1981, pp. 989-994.

round-trip efficiency in the range of 65%-75%.[281] An alternative fuel-free approach is isothermal CAES, which requires maintaining a constant temperature. Two R&D proposals are being supported by DOE efforts, however, few details are available on this technology.[282][283] In general, isothermal CAES requires maintaining highly efficient, responsive isothermal cycles with rapid heat transfer over a broad temperature range and it is still too early to gauge the merits of this approach. It is also unclear how this approach will affect part load efficiency or ramp rate capabilities, and the ability to serve multiple grid support functions.

Other Novel Approaches

There are several novel approaches to CAES cycles with limited active R&D efforts. One proposed configuration is aboveground CAES with air storage in high pressure piping.[284] Aboveground CAES has been proposed as an alternative air storage option for distributed (<10MW) short duration (< 3 hour) storage applications. While the capital cost of incremental storage capacity additions are significantly larger for such systems (~$200/kWh for aboveground CAES versus $2/kWh for salt dome storage), small, modular CAES systems without geologic siting constraints could be attractive for distributed, high-value applications.

Another approach for expanding the siting availability for CAES is to expand air storage into subsea environments. Underwater CAES concepts typically employ an anchored containment bag and exploit the buoyancy of air and the subsea pressure gradient to store high pressure air underwater.

Deployment Challenges

CAES offers many potential advantages: low capital cost, high efficiency, fast ramping capability, adaptability to many types of geologic storage, low fuel consumption, and a well-established operational record. Nevertheless, the deployment of CAES has been extremely slow and the past three decades have resulted in only a handful of utility-scale projects and test facilities. While in part this can be attributed to market conditions and regulatory barriers that have impacted the energy storage industry as a whole, there are technology-specific barriers as well.

A major barrier is the need to prove CAES operation in geologic formations other than domal salt. However, even after the demonstrations in individual locations mentioned previously, this does not demonstrate the universal applicability of CAES in porous rock.

The Electric Power Research Institute and others have shown that large fractions of the continental US have geology suitable for CAES, but these estimates only reflect the existence of salt, sandstone, and hard rock.[285] They do not take into account detailed geologic characteristics

---

[281] EPRI/DOE 2003, S. Succar & R.H. Williams 2008.

[282] General Compression, Inc "Fuel-Free, Ubiquitous, Compressed Air Energy Storage and Power Conditioning," fact sheet, 2010, . http://www.sandia.gov/ess/docs/pr_conferences/2010/slide1_marcus_gencomp.pdf

[283] D. Kepshire, "Isothermal Compressed Air Energy Storage," U.S. Department of Energy 2010 Energy Storage Systems Research Program Update Conference, Washington D.C., 2010, http://www.sandia.gov/ess/docs/pr_conferences/2010/kepshire_sustainx.pdf.

[284] EPRI/DOE, 2003.

[285] K. Allen, "CAES: The Underground Portion," *IEEE Transactions on Power Apparatus and Systems*, PAS-104, No. 4, 1985, pp. 809-12.

(continued...)

---

necessary to deem a site suitable for air storage and therefore do not provide a comprehensive analysis of deployment potential. While there is ample geology available for near-term deployment, an accurate determination of the total availability of geologic air storage will require extensive additional data. Site testing and characterization is certainly not a trivial matter; numerous screening criteria and extensive testing are required to determine a formation's adequacy for air storage. An extensive survey of CAES geologies including aquifer permeability, porosity, and cap rock characteristics is needed to accurately make the assessment on the deployment limits of CAES.

There are a number of other challenges associated with the underground formation and use. The development of solution mined caverns in domal or bedded salt requires water, and subsequent disposal of brine. Water use for solution mining is likely to be about 8 cubic meters of water for each cubic meter of salt excavated[286] or about 4.8 million cubic meters of fresh water withdrawals and brine management per 220-MW plant. Disposal of brine has been raised as a concern for some locations. Additional challenges related to the introduction of air into an underground geologic formation (e.g., oxidation and additional corrosion mechanisms) have been identified[287] and likewise cyclic loading of formations could result in greater material degradation than conventional natural gas storage operation[288] but it is not yet clear the extent to which these could limit CAES deployment potential. With regard to permitting the subsurface component, although it will depend on the type of geology and formation characteristics, air storage is similar in most respects to a conventional natural gas storage facility that have been routinely sited for many decades.[289]

Other deployment challenges associated with CAES are similar to other gas-fired power plants in terms of land,[290] water,[291] or other environmental impacts.[292] CAES requires little in the way of unique materials or labor requirements, and the above ground portion can be developed rapidly—the equipment required for CAES is very similar to conventional gas turbines, and the historical installation of gas turbines has often exceeded 8 GW per year.[293]

(...continued)

American Gas Association, *Survey of Underground Gas Storage Facilities in the United States and Canada*, Washington, DC, 2004.

[286] T. Smith, "Opportunities for Subsurface Compressed Air Energy Storage in New York State," Compressed Air Energy Storage (CAES) Scoping Workshop, Columbia University, October 21–22, 2008.

[287] Electric Power Research. Institute, *Compressed-Air Energy Storage: Pittsfield Aquifer Field Test*, Palo Alto, CA, 1990, p. 336.

[288] S.J Bauer and T.W. Pfeifle, "Potential Risks Associated with Underground CAES," Integrating Wind-Solar-CAES, 2nd Compressed Air Energy Storage (CAES) Conference & Workshop, Columbia University, 2010.

[289] Federal Energy Regulatory Commission, *Current State of and Issues Concerning Underground Natural Gas Storage*, Federal Energy Regulatory Commission, 2004.

[290] The land area estimate for one proposed CAES facility is about 140 m²/MW. Norton Energy Storage L.L.C., 2000.

[291] Cooling water is required during operation of the compressors, with one estimate of 2.5 million-3.0 million gallons per day for a 2700 MW facility. Assuming a 25% capacity factor (5913 GWh annual generation), this corresponds to about 0.2 gallons/kWh. See Ohio Power Siting Board, *In the Matter of the Application of Norton Energy Storage, LLC for a Certificate of Environmental Compatibility and Public Need for an Electric Power Generating Facility in Norton, Ohio*, Case No. 99-1626-EL-BGN, 2001.

[292] B.R Mehta, "Siting Compressed-Air Energy Storage Plants," Proceedings of the American Power Conference, Chicago, IL, Illinois Institute of Technology, 1990.

[293] U.S. Department of Energy, *Electricity Generating Capacity: Existing Electric Generating Units in the United States, 2008*, online table, 2011. http://www.eia.doe.gov/cneaf/electricity/page/capacity/capacity.html

Conclusions

CAES is a commercially available electricity storage technology with arguably the lowest cost for large-scale deployment. Its deployment is not limited by raw materials, calendar or cycle life, and can provide multiple services to the grid. The operational characteristics and benefits of CAES suggest that it could be a valuable addition to the generation mix and an important source of flexibility as the penetration of variable generation increases. The primary barrier to deployment is demonstration in widely available underground formations, including bedded salt, aquifers, and depleted gas wells. Perhaps the most important near-term R&D effort for CAES, in addition to demonstration in different formations, is a national screening to assess the regional and total potential for new development. While the available data suggests that CAES deployment could reach or exceed tens of gigawatts, additional analysis must be done to refine understanding of the geologic limits to CAES deployment. Additional R&D will be useful to identify alternative CAES configurations that reduce fuel use, improve performance, and use more standardized equipment.

Chapter 8: Electrochemical Capacitors

Overview

Electrochemical capacitors (ECs), including "supercapacitors" and "ultracapacitors," are devices that store energy in an electric field at the surface of an electrode.[294] Unlike traditional electrostatic capacitors, ECs use an electrolyte to shuttle ions between two working electrodes in a manner similar to batteries. Capacitors have among the fastest response time of any electricity storage device, and they are typically used in power-quality applications such as providing transient voltage stability. However, their low energy capacity has restricted their use to short time-duration applications with pulses lasting less than 40 seconds.

ECs are most commonly used in computer memory backup systems to bridge brief power interruptions. However, a target application of ECs is to act as an energy storage system for hybrid electric vehicles (HEVs) with low to moderate electric power requirements.[295] Compared to batteries, ECs have excellent cycle life and are well suited for cycling-intense applications such as use in transit buses and trains. A major research goal is to increase their energy density and thereby increase their usefulness in the grid and potentially in vehicle applications.[296] Capacitors have yet to see significant deployment in utility-scale or transportation applications, but there have been demonstration programs for both activities.[297]

Technology

Description

Traditional electrostatic capacitors store energy in an electric field between two electrodes. These devices are used extensively in consumer electronics, but their low energy density makes them unsuitable in utility or transportation applications. ECs differ from traditional capacitors in that they store energy in an "electric double-layer" that builds up at the surface of an electrode where it is wetted by an electrolyte. Similar to batteries, the electrolyte shuttles charged ions back and forth between two electrodes. Unlike batteries, capacitance at the electric double-layer serves as the primary energy storage mechanism of ECs. In contrast, batteries rely on a "faradaic" process for energy storage, meaning that the physical state of the electrode changes during charge/discharge, often limiting battery cycle life. **Figure 22** compares the various types of capacitors with batteries.

---

[294] There is some ambiguity within the industry regarding the name for capacitors with massive storage capability due to the many product names among manufacturers and the relative newness of the technology. "Electrochemical capacitors" is used herein as a generic term for this group of technologies See EPRI/DOE, 2003.

[295] J. Furukawa, T. Mangahara,, and L.T. Lam, "Development of the UltraBattery for Micro- and Medium-HEV Applications," Proceedings of the 214th Electrochemical Society Meeting, Honolulu, HI, October 12-17, 2008; J. Gonder, "Recent Analysis of UCAPs in Mild Hybrids," Presentation to the 6th Advanced Automotive Battery Conference, Baltimore, MD, May 17-19, 2006, http://www.nrel.gov/vehiclesandfuels/energystorage/pdfs/39731.pdf.

[296] I. Hadjipaschalis, A Poullikkas, and V. Efthimiou, "Overview of Current and Future Energy Storage Technologies for Electric Power Applications," *Renewable and Sustainable Energy Reviews*, No. 13, 2009, pp. 1513–1522.

[297] Examples include a 450 kW capacitor demonstration project for wind power smoothing at a water treatment plant in Palmdale CA, http://www.steab.org/docs/Energy_Storage_Program.pdf and a demonstration of the "UltraBattery" demonstration supported by ARRA.

Figure 22. Comparison of Various Capacitor and Battery Topologies

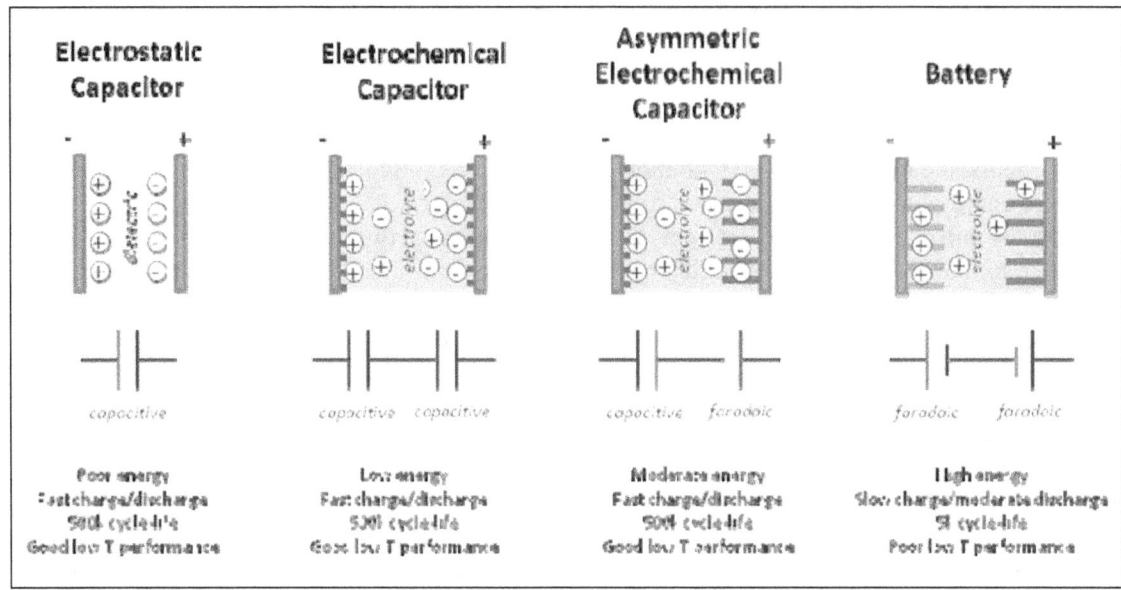

Source: K. Smith, National Renewable Energy Laboratory, 2011.

There are two general classes of ECs. The first is the electric double-layer capacitor which stores energy in the double-layer formed near the electrode surface. This type of capacitor includes the "supercapacitor" first introduced by NEC in 1978 and initially used to provide backup power for computer and appliance memory devices.[298] Although the energy is stored as charge like in a capacitor, special electrodes separated by an electrolyte are employed, similar to a battery. There are three types of electrode materials suitable for ECs—activated carbons with high surface area, conducting metal oxides, and conducting polymers. The high-surface-area carbon electrode material is the most common and least costly to manufacture. The electrolyte may be aqueous or organic.

The second type of EC more closely resembles a hybrid between a capacitor and battery. The capacitor-like electrode stores charge in the electric double-layer, while the battery-like electrode uses a faradaic mechanism. By employing some of the characteristics of batteries, they can greatly increase the energy density. These devices are referred to as "asymmetric" or "pseudo" capacitors and are currently under development.

Performance

ECs are capable of responding within a fraction of a second and are able to charge and discharge at high rates. As a result, they have a power density of ~5-10 kW/kg, much higher than most battery technologies which are typically less than 0.5 kW/kg.[299] However, EC energy density of 1-10 Wh/kg is at least 10 times lower than most battery types being considered for transportation applications.[300] As a result of these characteristics, a proposed application of capacitors is to place

---

[298] EPRI/DOE, 2003.

[299] J. Miller, and A. Burke, "Electrochemical Capacitors: Challenges and Opportunities for Real-world Applications," *ECS Interface*, Spring 2008. http://www.electrochem.org/dl/interface/spr/spr08/if_spr08.htm

[300] EPRI/DOE, 2003.

them in parallel with battery terminals and provide a current boost during periods of high demand, such as during vehicle acceleration. It is unclear though, whether the benefits of a parallel battery/EC system outweigh the additional cost of ECs; EV storage costs are predominantly driven by the useable energy requirement which dictates the size of the battery. Capacitors are designed to provide tens of thousands of charge/discharge cycles with limited or no performance degradation, with long lifetimes (on the order of 10 years) under continuous operation.[301] Accepted levels of performance degradation for ECs are 100% resistance growth and 20% to 30% capacity loss over 10 years and 500k cycles.

Unlike batteries, where voltage in the usable range is fairly constant, the voltage of an EC ranges from full voltage to zero volts. This can require additional electronics to boost EC power to the working voltage required, which makes control challenging. Provisions must be made to limit the current when charging a depleted EC. While a rapid discharge rate does not decrease the lifetime of an EC, research efforts are underway to extend EC discharge time. At present, EC rapid discharge rates are ideal for pulse power applications.

ECs perform well in cold environments, maintaining functionality down to -40°C. By comparison, lithium-ion batteries do not operate well below -10°C. ECs can charge and discharge with high turnaround efficiency—in excess of 95%.[302] Compared to batteries, this means that minimal thermal management is required, resulting in a simpler system overall. ECs are well suited to high-reliability applications in extreme environments which require frequent charge/discharge cycles with short bursts on the order of 10 seconds.

Commercial ECs have specific energies around 5 Wh/kg. Li-ion batteries, by comparison, have some 20 times more energy at 70-200 Wh/kg,[303] with perhaps one-fifth the power. Even in power applications, energy requirements commonly determine the size of the energy storage device. An example is a 50 kW uninterruptable power supply that must supply power for 2 minutes. The 50 kW power requirement would seem to indicate that just 10 kg of ECs are necessary. The 2 minute energy requirement, however, means that some 1700 Wh of energy storage is needed, or 340 kg of ECs. A Li-ion battery system is more likely to be power-constrained in this application, requiring 100-200 kg of batteries.

Figure 23 compares the energy and power density of ECs with various battery technologies. As demonstrated by the previous example, the inherent power-to-energy ratio of ECs dictates that they are best suited for short bursts of power from 0.1 to 40 second duration. By comparison, Li-ion batteries are best suited for charge/discharge operation ranging from 5 seconds to tens of hours.

---

[301] EPRI/DOE, 2003.

[302] DC-DC rating.

[303] J. Miller and A. Burke, 2008.

Figure 23. Comparison of Energy Storage Technologies

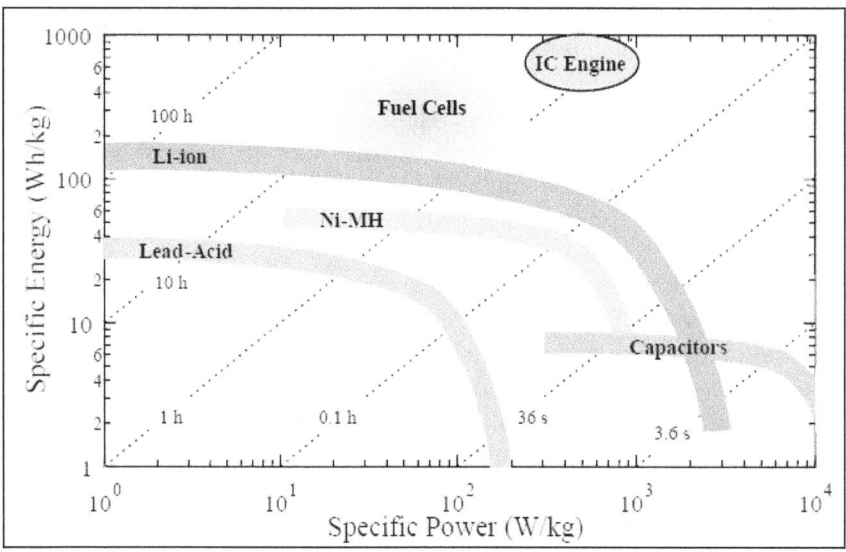

Source: V. Srinivasan "Batteries for Vehicular Applications—Present Status and Challenges," slide presentation, Lawrence Berkeley National Laboratory, April 22, 2009. http://www.ita.ucla.edu/news/presentations/ Srinivasan%20042209.pdf

Cost

Since 2000, the energy density of ECs has increased 250% and cost has fallen by 50%.[304] Costs of traditional symmetric ultracapacitors are as low as $10/kW and $17/Wh according to one manufacturer.[305] Allowing for 50% power degradation and 30% energy degradation over a 10 year, 500k cycle life, this manufacturer's usable power and energy costs are $20/kW and $24/Wh. These costs compare well with values listed in **Table 9**.

Table 9. Performance and Costs of Electrochemical Capacitors

Characteristic	Metric
Charge time	I second
Discharge time	I second
Cycle life	>500,000
Specific Energy (Wh/kg)	5
Specific Power (kW/kg)	5-10
Cycle efficiency (%)	<75% to >95%
Cost per unit energy ($/Wh)	$10-$20/Wh
Cost per unit power ($/kW)	$25-$50/kW

Source: J. Miller and A. Burke, 2008.

---

[304] M. Bolton, "Energy Storage Systems for Severe Duty Truck Applications," Presentation to the 5th International Symposium, Large EC Capacitor Technology and Application, Long Beach, CA, June 8-10, 2009.

[305] M. Leiber (Ioxus, Inc). "Falling Costs Heighten Appeal of Ultracapacitors," *Automotive Engineering International*, Jan. 11, 2011.

On a power basis, **Table 9** indicates that ECs are cheaper than Li-ion batteries. But as mentioned in the previous example, ECs often must be sized on an energy-basis. On an energy-basis, today's ECs initially do not appear to be cost competitive with Li-ion batteries, although this changes when considering applications requiring frequent cycling. Commercial ECs can tolerate 500,000 full depth-of-discharge (DoD) cycles. The cycle life of Li-ion batteries is highly dependent on how deeply the battery is discharged each cycle. For example, a graphite/nickelate Li-ion battery might last for 3,000 cycles at 80% DoD, and 500,000 cycles at 4% DoD.[306]

As a general rule, for applications with frequent cycling, ECs are cheaper than Li-ion batteries. **Figure 24** compares the cost of ECs and Li-ion batteries per unit energy throughput (calculated as device energy-specific cost divided by expected cycle-life times DoD). For Li-ion batteries, the expected cycle-life is dependent on DoD. For ECs, the cycle-life is assumed to be 500,000 cycles, independent of DoD. EC life is assumed to be 10 years. As the figure shows, ECs are the cheaper alternative to Li-ion batteries only for cycling-intense applications requiring 100 or more cycles per day. This conclusion is, of course, highly dependent on cost and life assumptions for the two technologies.

Figure 24. Comparison of Cost per Energy Throughput for Li-Ion Batteries and ECs

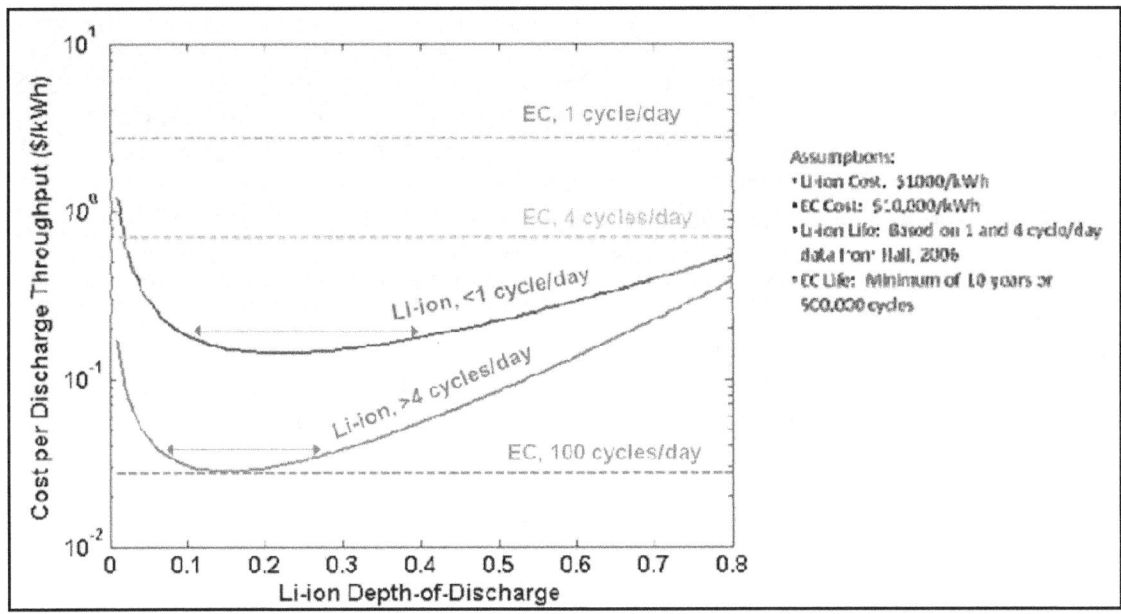

Source: K. Smith, National Renewable Energy Laboratory, 2011.

Note: The figure uses two different logarithmic relationships from J.C. Hall et al. fit to data over a range of cycling duties.

Research and Development

Current EC research primarily targets reducing material and device cost and increasing energy density without sacrificing power and life. Specific areas of research and development include:

---

[306] J.C. Hall, et al., "Decay Processes and Life Predictions for Lithium Ion Satellite Cells," Paper No. AIAA 2006-4078, 4th International Energy Conversion Engineering Conference and Exhibit, San Diego, CA, June 26-29, 2006.

- Reducing component and finished electrode material manufacturing costs,

- Increasing the capacitance of electrodes by increasing surface area and tailoring the pore size and shape,

- Finding electrolytes capable of voltages beyond 2.7V, preferably with less toxicity, and

- Optimizing asymmetric ECs, with potential to increase energy density to 8 times that of standard ECs.[307]

Advanced Carbons

Carbon remains the preferred material for EC electrodes as it is non-reactive in most electrolytes. Carbon can be derived from a variety of materials and its structure is tunable during manufacturing, allowing the designer to control surface area, pore size and pore volume.[308] While the cost of raw carbon may be low, highly purified finished carbon is generally expensive. However, carbon electrodes have the potential to cost less in the future. Nanotube and graphene structures are also under investigation as possible EC electrode materials.

Increasing Capacitance

The capacitive effect, responsible for storing energy, occurs at the electrode/electrolyte interface. Finished carbons have tailored pore shapes and sizes, creating high specific surface area with a large working capacitance. Typical surface areas are 1000-2000 square meters per gram of material. For a fixed amount of surface area, two effects contribute to EC double layer capacitance, (1) the space charge layer and (2) the Helmholtz layer.[309] A suitable space charge layer requires electrode wall structures with a thickness greater than about 1 nm. In order for charged ions to build up at the Helmholtz layer, electrode pores must be large enough for the ion to fit. Until recently, it was thought that pore sizes needed to be greater than 2 nm to accommodate the charged ion as well as its accompanying electrolyte solvent sheath. However, it was recently discovered that smaller pores might also be accessible, with greatly increased capacitance.[310]

The impact is that far greater capacitance may be available from tailored electrodes than previously thought possible. Better physical understanding of capacitive behavior is expected to lead to new materials with greater energy density.[311]

---

[307] An overview of EC research is provided in: *Interface,* Electrochemical Society, Vol 17, No. 1, Spring 2008. http://www.electrochem.org/dl/interface/spr/spr08/if_spr08.htm

[308] R. Brodd, "Overview of Electrochemical Capacitors," Presentation to the 27th International Battery Seminar and Exhibit. Mar. 16-18, 2010.

[309] H.J. Gerischer, "An Interpretation of the Double Layer Capacity of Graphite Electrodes in Relation to the Density of States at the Fermi Level," Journal of Physical Chemistry, No. 89, 1985, p. 4249.

[310] J. Chmiola, et al., "Anomalous Increase in Carbon Capacitance at Pore Sizes Less Than 1 Nanometer," *Science,* No. 313, 2006, p. 1760.

[311] J.S. Huang, B.G. Sumpter, and V. Meunier, "Theoretical Model for Nanoporous Carbon Supercapacitors,"*Angewandte Chemie, International Edition,* Vol. 47, No. 3, 2008, pp. 520-524.

High Voltage/Alternative Electrolytes

Greater energy density is possible by raising the working voltage of the EC; however, conventional electrolytes break down at voltage which is too high. Common electrolytes include acetonitrile and propylene carbonate—each permit high operating voltages up to 2.7V. Electrodes may also be a limiting factor at high voltages as electrode shrinking and swelling occurs with cycling, inducing stress that limits cycle life,[312] similar to battery electrodes. The calendar life of EC devices is presently limited by electrolyte degradation which gradually builds up pressure inside the EC case. EC life beyond 10 years is uncertain, though possible, as long as the device does not accumulate too much time at high voltages and/or temperatures.

Asymmetric Designs for Increased Energy

Asymmetric designs replace the positive electrode in carbon-carbon ECs with a faradaic, battery-type electrode such as $NiOOH$, $MnOOH$, PbO^2, and $Li_4Ti_5O_{12}$. Introducing a high-capacity electrode reduces mass and volume and allows for a flatter operating voltage. Energy density may increase by eight times.[313] Replacing a capacitive electrode with a faradaic one, however, introduces material volume expansion and contraction issues, accompanied by mechanical stress and electrode fracture that can limit cycle life. Nonetheless, there is sufficient middle ground between traditional battery performance with power-to-energy ratio of ~5 hr^{-1} and EC performance with power-to-energy ratio of ~1000 hr^{-1} to warrant further research and optimization of these hybrid devices. A number of promising configurations and electrochemical couples are under investigation.[314]

Deployment Challenges

ECs have deployment challenges similar to several battery types. As with batteries, capacitors present potentially lethal voltage levels. For utility (grid) applications, this presents very little incremental risk, but presents challenges for transportation applications. Aqueous electrolytes may contain hazardous materials including potassium hydroxide and methyl cyanide. Furthermore, certain electrolytes are flammable, such as acetonitrile, which releases hydrogen cyanide when burned.[315][316] This may provide limited risk in grid applications, where there is lower risk of release, and the expectation is that installation and maintenance would be performed by trained personnel only. As with batteries, ECs must be properly disposed or recycled at end-of-life.

The majority of materials in current ECs include common materials such as carbon, nickel, steel, aluminum, and a variety of plastics. Advanced asymmetric ECs would use several materials used

---

[312] R. Kötz, et al., "In-situ Monitoring of EDLCs Operation by Physio-Chemical Techniques," Presentation to the 5th International Symposium, Large EC Capacitor Technology and Application, Long Beach, CA, June 8-10, 2009.

[313] R. Brodd, "An Overview of Electrochemical Capacitors," Presentation to the 27th International Battery Seminar and Exhibition, Ft Lauderdale, Fl, March 16-18, 2010.

[314] K. Naoi and P. Simon, "New Materials and New Configurations for Advanced Electrochemical Capacitors," *Interface*, Electrochemical Society, Vol 17, No. 1, Spring 2008. http://www.electrochem.org/dl/interface/spr/spr08/if_spr08.htm

[315] EPRI/DOE, 2003.

[316] K. Rechenberg and M. Meinert, "Requirements on DLC Energy Storage Units for Rolling Stock," Presentation to the 5th International Symposium, Large EC Capacitor Technology and Application, Long Beach, CA, June 8-10, 2009.

---

in advanced batteries, such as lithium and vanadium. It is difficult to estimate the total material requirements, but they would unlikely be greater than those for batteries, and this requirement must be placed in the context that the target applications for capacitors are those with limited actual energy capacity.

Conclusions

In the near term, EC systems are likely limited to power-related (rapid discharge) applications for both grid and transportation applications. Currently, ECs have high power density, and are cost-competitive for certain applications where discharge time is measured in seconds. The low energy density and high cost per unit of energy stored makes ECs currently uncompetitive for energy applications where discharge times of minutes or more are required. Energy density will need to increase by at least an order of magnitude for capacitors to compete against batteries for electricity storage applications in transportation. ECs are already well suited for niche applications such as trains and buses, requiring high power, low temperatures, and/or high cycle-life. These early markets continue to decrease EC manufacturing costs.

Chapter 9: Pumped Hydro Storage

Overview

Pumped hydro storage (PHS) stores energy by pumping water from a lower-level reservoir (e.g., a lake or river) to a higher-elevation reservoir via an underground tunnel.[317] During periods of high electricity demand, the water is released to the lower reservoir to turn turbines to generate electricity, similar to the way in which conventional hydropower plants generate electricity. Many existing PHS plants store eight hours or more of energy, making them useful for load leveling, and providing firm capacity. PHS can also ramp rapidly, making it useful for load following and providing ancillary services including contingency spinning reserves and frequency regulation.[318]

Pumped hydro is the only electricity storage technology deployed on a gigawatt scale worldwide. In the United States, about 22 GW is deployed at 39 sites,[319] while global capacity is over 127 GW.[320] U.S. capacity was largely built during the 1970s and 1980s largely in response to market conditions in the 1970s as discussed in Chapter 2. **Figure 25** illustrates the capacity of PHS in the United States There has been no large-scale PHS development in the United States since 1995,[321] although development has continued in Europe and Asia.[322] Lack of continued development of PHS projects domestically has been attributed to a number of factors, including the availability of low-cost natural gas along with increasing regulatory, environmental, and siting challenges.[323]

---

[317] There are several names and acronyms for pumped storage. They include pumped-storage hydro (PSH) pumped hydro storage (PHS), hydro-pumped storage (HPS) and pumped hydro energy storage (PHES).

[318] J. Phillips, "Pumped Storage in a Deregulated Environment," *International Journal on Hydropower & Dams*, Vol. 7, No. 1, 2000, pp. 32-35.

[319] Estimates for the total PHS installed capacity in the United States ranges from 20-22 GW, partially due to different plant ratings. For example, the EIA lists the total nameplate capacity of PHS as of 2010 at 20.5 GW, while the summer capacity is listed at 22.2 GW. U.S. Department of Energy, "Existing Generating Unit in the United States by State and Energy Source, 2010," online database, 2011. http://www.eia.gov/electricity/capacity/xls/existing_gen_units_2010.xls

[320] E.A. Ingram, "Worldwide Pumped Storage Activity," *Hydro Review Worldwide*, Vol. 18, No. 4, September 2010.

[321] There is one small (40 MW) PHS facility which began operating in 2011. San Diego County Water Authority, "Lake Hodges Project Begins Pumped Storage and Power Generation Operations," press release, September 14, 2011.

[322] J.P. Deane, B.P. Ó Gallachóir, and E.J McKeogh, "Techno-Economic Review of Existing and New Pumped Hydro Energy Storage Plant," *Renewable and Sustainable Energy Reviews*, Vol. 14, No. 4, May 2010, pp. 1293-1302.

[323] P. Denholm, E. Ela, B. Kirby, and M. Milligan, *The Role of Energy Storage with Renewable Electricity Generation*, NREL/TP-6A2-47187, National Renewable Energy Laboratory, Golden, CO, 2010.

Figure 25. Capacity of PHS in United States, 1956–2003

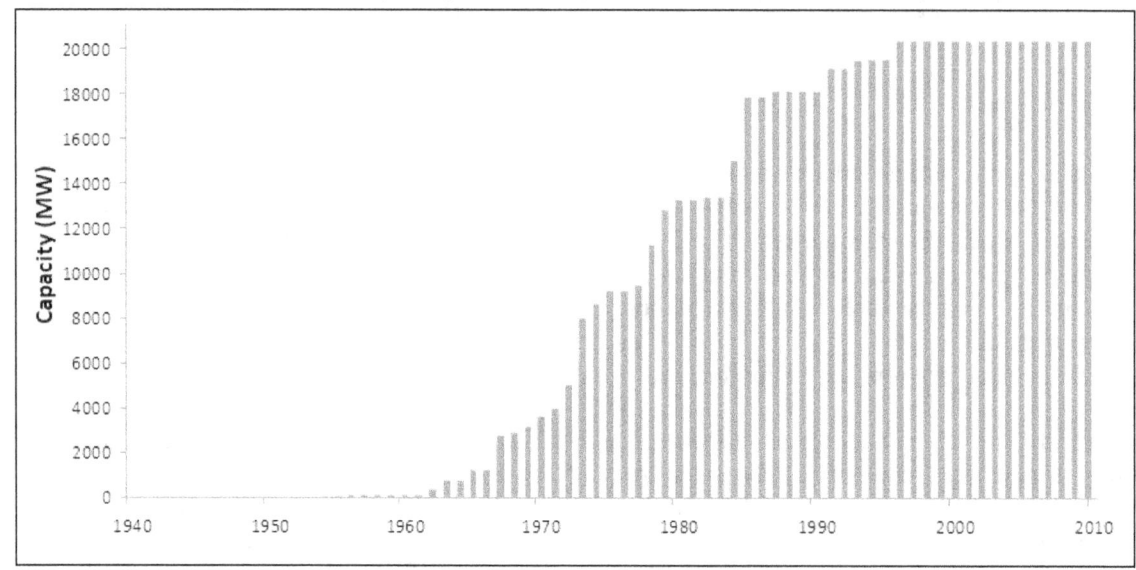

Source: Derived from U.S. Department of Energy, "Existing Generating Units in the United States by State and Energy Source, 2010," online database, 2011.

Although domestic PHS capacity has been relatively static for the last 15 years, developers are showing a renewed interest in building PHS projects in the United States. As of December 2011, the Federal Energy Regulatory Commission (FERC), which regulates PHS projects, issued preliminary permits for 45 new plants, representing about 35 GW of capacity.[324] The capacity of proposed plants (including those with issued and pending preliminary permits) exceeds 40 GW. **Figure 26** is a map of existing PHS sites as well as those with a FERC preliminary permit.

---

[324] Federal Energy Regulatory Commission (FERC), "All Issued Preliminary Permits," internet database, December 2011. http://www.ferc.gov/industries/hydropower/gen-info/licensing/issued-pre-permits.xls

---

Figure 26. Existing and Proposed PHS Facilities in the United States

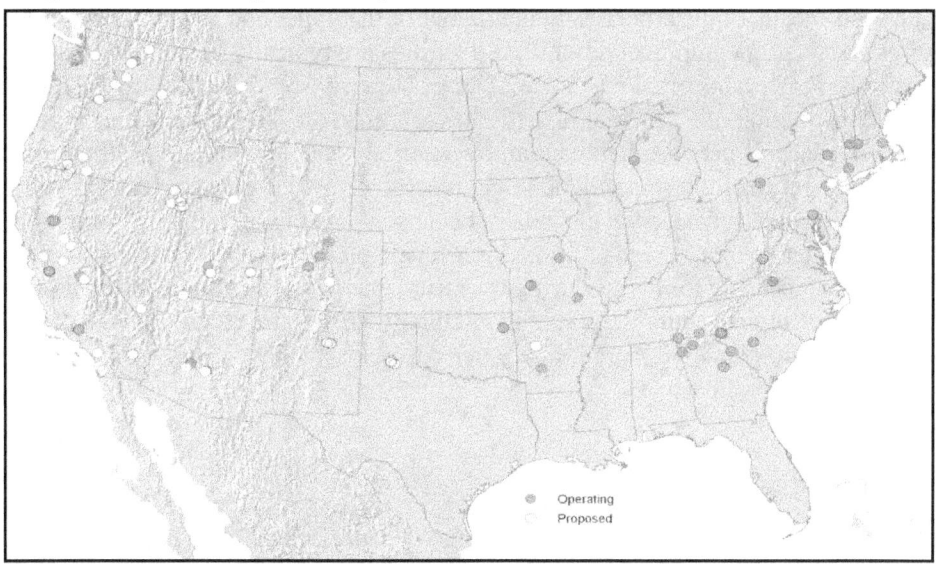

Source: D. Heimiller, National Renewable Energy Laboratory.

Technology

Description

PHS consists of two reservoirs connected by an underground shaft and a powerhouse containing turbines and electrical equipment. PHS relies on either reversible pump-turbine motor-generator units or on separate motor/pumps and turbine/generators. Reversible units operate as a motor and pump in the "pumping" mode, and as a turbine and generator in the "generating" mode. The great majority of U.S. plants have multiple reversible pump-turbines. **Figure 27** shows a representative configuration of a PHS plant.

Figure 27. Pumped-Storage Hydropower Plant Configuration

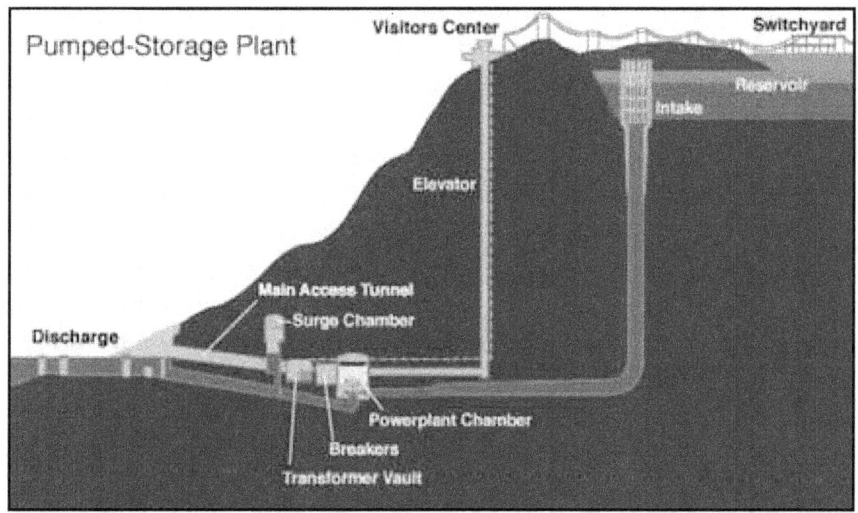

Source: Tennessee Valley Authority, "The Mountaintop Marvel," web page, 2011. http://www.tva.gov/heritage/mountaintop/index.htm

---

The total amount of energy stored in a PHS facility is the product of the volume of the upper reservoir and the "head" or difference in height between the upper and lower reservoir. Siting PHS requires suitable topographical relief. PHS facilities are typically large and located in fairly remote locations. PHS plants often make use of an existing river or lake, avoiding the need for—and cost of—construction of a separate (usually lower) reservoir. This is called an "open cycle" PHS plant. In instances where a suitable natural water body is not available for use as one of the reservoirs, both the upper reservoir and the lower reservoir must be constructed. This type of construction is known as a "closed-cycle" plant, because it has minimal interaction with natural water bodies. A water source is needed for a closed-cycle plant to provide water to initially fill the reservoir and to compensate for losses during operation due to leakage and evaporation. Nearby rivers or streams are typical sources; treated municipality grey water or groundwater (wells) can also be used. Of the 40 plants with preliminary permits at least 9 have proposed closed-cycle PHS plants, exceeding 9 GW of capacity.[325]

Performance

Existing PHS installations in the United States range in capacity from less than 50 MW to 2,800 MW with typical energy capacities in the range of 8-15 hours of full discharge.[326] Most PHS plants can ramp rapidly while generating and are often used for ancillary services. Some modern PHS plants can also rapidly change the rate of pumping. The greatest limitation in older PHS plants is the time required to switch between pumping and generation which can be up to 30 minutes.

The round-trip efficiency of a PHS plant depends largely on the type of pump/turbine system and the head. PHS plants use and generate AC electricity, avoiding the conversion losses (and costly power electronics) associated with technologies that store or generate DC electricity—such as batteries. Typical AC-AC efficiencies for U.S. plants are in the range of 65%-80%.[327] There has been a trend toward increased efficiencies. Proposed PHS plants have expected efficiencies exceeding 80%. **Figure 28** illustrates the efficiency tend for U.S. facilities. Transmission requirements and associated losses may slightly reduce the effective efficiency of PHS.[328] There is little loss of performance due to age or throughput. Plants are upgraded through efficiency improvements and life extension on a project-by-project basis, and most U.S. projects have been modernized through runner (turbine) replacements, generator rewinds, control system upgrades, and other improvements.[329] For example, the New York Power Authority upgraded its Belenheim-Gilboa plant to increase its operating range from 203-260 MW to 140-290 MW.[330] Lifetimes of

---

[325] FERC, 2011.

[326] U.S. Department of Energy, "Existing Generating Units in the United States by State and Energy Source, 2010," online database, 2011.

[327] Task Committee on Pumped Storage of the Hydropower Committee of the Energy Division of the American Society of Civil Engineers (ASCE), *Compendium of Pumped Storage Plants in the United States*, American Society of Civil Engineers, New York, NY, 1993.

[328] P. Denholm, and G.L. Kulcinski, "Life-Cycle Energy Requirements and Greenhouse Gas Emissions from Large-Scale Energy Storage Systems," *Energy Conversion and Management*, No. 45, 2004, pp. 2153-2172.

[329] A. Ferreira "Sixteen Years Operating and Maintenance Experience of the 1080 MW Northfield Mountain Pumped Storage Plant" and B.E. Sadden "Maintenance of Pumped Storage Plants," *Pumped Storage*, Institution of Civil Engineers, London, England, 1990.

[330]K Tani and H. Okimura, "Performance Improvement of Pump-Turbine for Large Capacity Pumped Storage Power Plant in US," *Hitachi Review*, Vol. 58 No. 5, October 2009.

PHS plants can exceed 60 years; the Rocky River Plant in Connecticut has operated since 1928.[331]

Figure 28. Historical Efficiencies of PHS Plants in United States

Source: National Renewable Energy Laboratory, 2011. Compiled from several sources including ASCE 1993, FERC "Form 1" filings and Department of Energy databases.

Having multiple units per plant allows for scheduling maintenance on one unit while keeping the other units available, typically minimizing effects on overall plant availability. Existing PHS facilities in the United States have high availability and few forced outages.[332] New PHS deployments in the United States would likely include variable speed (also referred to as "adjustable speed") operation. This technology has not yet been applied in a major U.S. installation, but has been used in several international plants.[333] Among the benefits of variable speed operation are faster response to grid requirements, higher efficiencies, ability to accommodate greater ranges of "head," and wider unit and plant operating ranges (i.e., an ability to operate with a lower minimum load in megawatts).

Cost

The cost of new PHS plants will vary. Since large-scale PHS has not been built in the United States in some time, the cost of the next plant is somewhat uncertain. The geotechnical and geological characteristics and complexity of a site are major factors in PHS development costs. Typically, the largest costs are for development of a project's upper and lower reservoirs and for underground components. **Figure 29** provides historical cost data for U.S. PHS plants. There is a general trend toward increasing costs, with the last three plants constructed costing over

---

[331] EIA, "Existing Electric Generating Units in the United States."

[332] Detailed availability and outage statistics are available via the Generating Availability Data System (GADS) from the North American Electric Reliability Corporation, http://www.nerc.com/page.php?cid=4|43.

[333] M. Yasuda, "Enhancing Ancillary Services to Make Pumped Storage More Competitive," *International Journal on Hydropower & Dams*, Vol. 7, No. 1, 2000, pp. 36-42.

$1000/kW. The cost for PHS is typically reported only in terms of cost per unit of capacity ($/kW) and includes both the energy component and power component.

Figure 29. Installed Cost of PHS Plants in United States

(current dollars)

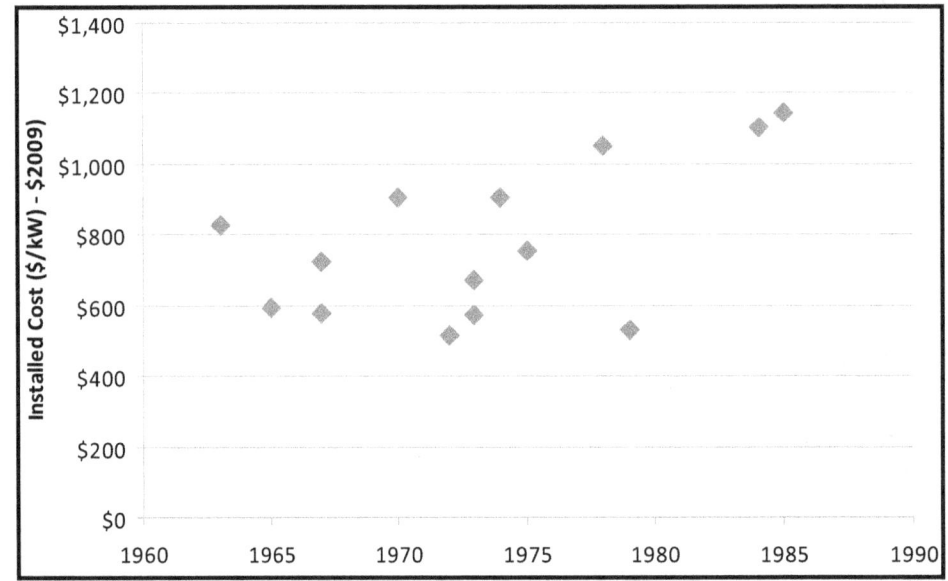

Source: National Renewable Energy Laboratory, 2011. Compiled from several sources including ASCE 1993, FERC "Form 1" filings and Department of Energy databases.

A number of projects have been completed worldwide in the last decade. There are also a significant number of proposed plants both in the United States and internationally. **Table 10** lists several recently completed plants in Europe and proposed plants in the United States, along with capital costs ($/kW) adjusted to $2009. There are many proposed plants in Europe, with costs estimated in the range of $700/kW to over $3000/kW.

Table 10. Recently Completed or Proposed PHS Plants

Location	Plant Name	Capacity (MW)	$/kW (2009)	Date of Completion
U.S.				
California	Eagle Mountain	1300	1019	Proposed
California	Iowa Hills PS	400	1344	Proposed
California	Lake Elsinore	500	1500	Proposed
California	Red Mountain	900	1900-2100	Proposed
Utah	North Eden PS	700	1011	Proposed
Utah	Parker Knoll PS	800	1215	Proposed
Austria	Feldsee	140	750	2009
Austria	Reisseck_II	430	1091	2008
Germany	Goldisthal	1060	1321	2003
Slovenia	Avce	180	711	2009

Sources: National Renewable Energy Laboratory compilation, 2010. Europe costs from J.P. Deane et al., 2010. United States costs are from various public sources including Northwest Wind Integration Forum, Pumped Hydro Storage Workshop, Portland, OR, October 17, 2008.

Many of the lower-cost projects listed in the table require only the construction of one reservoir (typically the upper reservoir) and use an existing body of water, abandoned mine, or other existing formation for the lower reservoir. New construction requiring an upper reservoir will raise costs significantly.[334] Engineering estimates of more "generic" PHS plants are often higher than the costs cited in **Table 10**. Examples include estimates from the Electric Power Research Institute ($2100-$4000/kW)[335] and R.W. Beck ($5595/kW).[336] The cost impact of using variable speed equipment in new plants is expected in the range of $50/kW (assuming a base cost of $1000/kW) to $200/kW (over a base cost of $1900/kW).[337]

Research and Development

PHS is considered a mature technology, so there is little active R&D dedicated to it. However, incremental improvements in efficiency are possible, and the flexibility of existing and future plants may be improved using variable speed drive technologies.[338]

---

[334] A more comprehensive discussion of recent and projected future costs is provided by J.P. Deane et al., 2010.

[335] $2,100 in D. Rastler, *Overview of Electric Energy Storage Options for the Electric Enterprises*, Electric Power Research Institute, 2009; $2,500-$4,000 in D. Rastler, "New Demand for Energy Storage," *Electric Perspectives*, Edison Electric Institute, September/October 2008.

[336] This cost is based on a 250 MW plant, with a basic breakdown of cost components provided. Further details of the type of PSH plant are unclear, but it is presumably a "greenfield" site requiring extensive development of both reservoirs. See Energy Information Administration, *Updated Capital Cost Estimates for Electricity Generation Plants*, November 2010.

[337] The lower estimate is from S.M. Schoenung and W.V. Hassenzahl, "Long- vs. Short-term Energy Storage Technologies Analysis—A Life-Cycle Cost Study," Sandia National Laboratories, 2003. The Higher estimate is from A. Petersen, "Red Mountain Bar Pumped Storage Project," Presentation to the Northwest Wind Integration Forum, Pumped Hydro Storage Workshop, Portland, OR, October 17 2008. Both estimates are in uninflated dollars.

[338] A meeting of PHS stakeholders held in 2010 identified 10 key issues associated with deployment. Only one item on the list was technology related (demonstration of variable speed equipment). Oak Ridge National Laboratory, *Pumped* (continued...)

Although not technical, one potential research effort considered important by some would be a national screening of potential PHS sites. PHS suffers from the perception that there are no available sites for new development, notwithstanding that the capacity of proposed plants exceeds the installed capacity of PHS in the United States.[339] Although there is no single comprehensive estimate of PHS potential, older studies indicate the availability of hundreds of conventional PHS sites and over 1000 GW of potential capacity in just 6 western states.[340] Over 100 GW of potential has been identified in the eastern states.[341] These older assessments include some areas that would be very difficult (or impossible) to develop based on current environmental or other restrictions. Efforts are underway at DOE national labs to perform resource assessment for new PHS development.[342]

Additional R&D could evaluate unconventional PHS development. PHS requires suitable topographical relief, but there is a noticeable lack of existing or proposed sites in much of the country, particularly in the Midwest and Texas (both areas with excellent wind resources). There are several proposals to extend PHS deployment in areas without traditional PHS geography. There are also proposals to decrease the footprint of PHS, which would have the added benefit of reducing its environmental impact and corresponding opposition. The most common proposal is to use an underground formation such as a natural aquifer or mined cavern for the lower reservoir. There was extensive discussion of this type of configuration, including a large number of design studies and proposals for underground pumped hydro, during the build-out period of conventional PHS during the 1970s and 1980s.[343] Of the proposed plants with preliminary permits, several use an underground formation, but the majority use an above ground reservoir. This is likely due to cost, but updated cost estimates would be valuable to assess the feasibility of underground PHS. Other approaches to PHS include saltwater PHS facilities in coastal regions, where the ocean is the lower body.[344]

(...continued)

Storage Hydropower, 2010, http://www.esd.ornl.gov/WindWaterPower/PumpedStorageSummitSummarySep2010.pdf.

[339] C-J. Yang and R.B. Jackson., "Opportunities and Barriers to Pumped-hydro Energy Storage in the United States," *Renewable and Sustainable Energy Reviews*, No. 15, 2011, pp. 839–844.

[340] This represents 155 potential sites with 341 GW of capacity in Arizona, California, Nevada, and Utah, as well as 670 GW. Harza Engineering Co. and U.S. Department of Energy, *Underground Pumped Hydro Storage and Compressed Air Energy Storage: An Analysis of Regional Markets and Development Potential,* Argonne National Laoboratory, 1977. This report is also summarized in A. E. Allen, "Potential for Conventional and Underground Pumped-Storage," *IEE Transactions on Power Apparatus and Systems*, Vol. PAS-96, No. 3, May/June 1977.

[341] Dames and Moore, *An Assessment of Hydroelectric Pumped Storage*, IWR 82-H-10, prepared for the U.S. Army Engineer Institute for Water Resources, 1981.

[342] L. Rogers, T. Key, and P. March, "Quantifying the Value of Hydropower in the Electric Grid," presentation to the 4th International Conference on Integration of Renewable and Distributed Energy Resources, Albuquerque, NM, December 6-10, 2010.

[343] A fairly large body of literature exists, mostly published between 1975 and 1985. For example, see Electric Power Research Institute, *Preliminary Design Study of Underground Pumped Hydro and Compressed-Air Energy Storage in Hard* Rock, EPRI-EM-1589, 1981. A brief literature review of this topic is provided by G. Martin, "Aquifer Underground Pumped Hydro," Colorado Energy Research Institute, June 30, 2007 and W.F. Pickard, A.Q. Shen, and N.J. Hansing, "Parking the Power: Strategies and Physical Limitations for Bulk Energy Storage in Supply-Demand Matching on a Grid Whose Input Power is Provided by Intermittent Sources, " *Renewable and Sustainable Energy Reviews*, Vol. 13, No. 8, October 2009, pp. 1934-1945.

[344] This approach has been demonstrated in Japan. H. Tanaka, H., "The Role of Pumped-storage in the 21st Century," *International Journal on Hydropower & Dams*, Vol. 7, No. 1, 2000.

Another pumped hydro storage method has recently been proposed that works by hydraulically moving a large cylindrical weight inside a vertical underground pipe.[345] A cylindrical weight sitting on a column of water is raised like a piston to store electricity and lowered during discharge. The total amount of energy that can be stored is proportional to the mass of the piston and the length of the vertical column. Since these are excavated systems, they could theoretically be sited anywhere with suitable geology and could have a relatively small footprint (about 0.3 GW/acre).[346] The proposed roundtrip efficiency of this device is likely to be comparable to conventional PHS systems, about 75%-80%.

Deployment Challenges

There are a number of barriers to further development of PHS. While a major barrier appears to be capital cost, an element of this cost is the long construction time, and associated risks and uncertainty, especially under changing market conditions and structures. Federal Energy Regulatory Commission permitting alone requires about five years.[347] State and local application and permitting can add to this time. Construction times vary, with one recent estimate of 4-5 years. This results in a 10-12 year construction time for new PHS based on current schedules.

Permitting and construction times, and associated costs, can increase due to siting opposition and environmental regulations. PHS development on existing streams can affect water quality and ecosystems as with any other hydro project.[348] Environmental concerns have prevented or greatly delayed construction of many proposed projects, and even delayed or prevented operation of completed projects. For example, one project in Georgia was completed in 1988, but did not operate until 2002 due to environmental concerns.[349] Another project in Missouri was completed in 1982, but has never been used due to fish kills.[350] Finally, there is significant opposition to new PHS development based simply on the amount of land area flooded for the upper and lower reservoirs. The total flooded area of three of the more recently constructed large PHS plants in the United States (the Bad Creek Hydroelectric Station in South Carolina, the Balsam Meadow Pumped Storage Project in California, and the Bath County Pumped Storage Station in Virginia) is in the range of 1,200 m^2/MW-1,500 m^2/MW.[351] Older PHS facilities with constructed upper and lower reservoirs can have flooded areas that exceed 4,000 m^2/MW. New plants are more likely to have land use requirements towards the lower range, such as the proposed Eagle Mountain and

---

[345] P. Reynolds, "A Weighting Game," *Water Power and Dam Construction,* March 2010, http://www.launchpnt.com/ Documents/Gravity-Power-Module-Article-IWPDC-March-2010.pdf.

[346] E. Wesoff, "Gravity Power's New Take on Pumped-Hydro Energy Storage," GreentechMedia, November 9, 2010. http://www.greentechmedia.com/articles/read/Gravity-Power.

[347] D. M. Adamson, "Realizing New Pumped-Storage Potential through Effective Policies," *Hydro Review*, April 2009, pp. 28–30.

[348] For examples, see J.P. Clugston (editor), *Proceedings of the Clemson Workshop on Environmental Impacts of Pumped Storage Hydroelectric Operations,* U.S. Fish and Wildlife Service, 1980; P.L. Strauss, "Pumped Storage, the Environment and Mitigation," proceedings of Waterpower '91: A new View of Hydro Resources, New York, NY, 1991; G.M. Simmons, Jr. and S.E. Neff, "The Effect of Pumped-Storage Reservoir Operation on Biological Productivity and Water Quality," Water Resources Research Center, Virginia Polytechnic Institute, 1969.

[349] C.-J. Yang and R.B. Jackson, "Opportunities and Barriers to Pumped-Hydro Energy Storage in the United States," *Renewable and Sustainable Energy Reviews,* Vol. 15, 2011, pp. 839-844.

[350] General Accounting Office, *Power Marketing Administrations: Cost Recovery, Financing, and Comparison to Nonfederal Utilities*, GAO/AIMD-96-145, September 1996, p. 34.

[351] ASCE 1993.

Iowa Hill plants, with flooded area requirements of about 1,100 m^2/MW.[352] Since PHS facilities are generally large and in remote, they typically require new high-voltage transmission, which adds additional siting challenges.[353]

An approach to minimizing the environmental impact (and potential opposition) to new PHS construction is to reduce or eliminate its impact on existing bodies of water. Most existing U.S. PHS plants are "open-cycle" plants; that is, they use an existing water body, usually the lower reservoir, for one of their reservoirs. However, "closed-cycle" plants—plants where both lower and upper reservoirs are constructed—will likely become more prevalent in the future because they minimize environmental effects as they do not interact with natural water bodies and they have little or no impact to aquatic life. Closed cycle plants can use existing features including abandoned mines to minimize development costs. A water source is needed for a closed-cycle plant to provide water to initially fill the reservoir and to compensate for losses during operation due to leakage and evaporation. Some proposed plants will use groundwater and at least one facility has proposed to use recycled wastewater, which could be a significant opportunity for other new PHS facilities as well.[354] Closed-cycle PHS plants could be candidates for a streamlined FERC permitting process given their lack of interaction with any active body of water.[355] This could reduce licensing and construction times to eight years, reducing investor risks.

Safety risks associated with PHS are similar to those of other hydro projects. The biggest risk is dam failure and flooding. There is one example of failure (resulting in property damage and injuries) at a U.S. PHS facility, which occurred at the Taum Sauk plant in 2005.[356] The failure was due to the upper reservoir being overfilled and subsequently being breached. The resulting flood destroyed one house and caused damage to a local park. PHS requires little if any toxic, rare, or foreign-sourced materials. There are no operational emissions. "Life-cycle" greenhouse gas emissions due to construction and operation are relatively low.[357] This includes emissions from decomposition of vegetation on land flooded by new PHS reservoirs, which are relatively small, especially for U.S. sites.[358]

---

[352] G. Tam, "Eagle Mountain Hydro Electric Pumped Storage Project," presentation to the Northwest Wind Integration Forum, Portland, OR, October 17, 2008, http://www.nwcouncil.org/energy/wind/meetings/2008/10/GilTam.pdf; D. Hanson, "Iowa Hill Pumped Storage Development Project Description," presentation to the Northwest Public Power Association Conference, April 6, 2011. http://www.nwppa.org/web/presentations/2011_EandO/Hansen_Iowa_Hill_New_Project.pdf

[353] The proposed Lake Elsinore Advance Pumped Storage project requires about 32 miles of new 500 kV transmission routed around the San Mateo Canyon Wilderness. Federal Energy Regulatory Commission, *Final Environmental Impact Statement Lake Elsinore Advanced Pumped Storage Project*, FERC/FEIS – 0191F, January 2007.

[354] C.-J. Yang and R.B. Jackson, 2011.

[355] D.M. Adamson, April 2009.

[356] Federal Energy Regulatory Commission, *Report of Findings on the Overtopping and Embankment Breach of the Upper Dam - Taum Sauk Pumped Storage Project*, FERC No. 2277, April 28 2006.

[357] P. Denholm and G.L. Kulcinski, 2004.

[358] L.P. Rosa and M.A. dos Santos, "Certainty & Uncertainty in the Science of Greenhouse Gas Emissions from Hydroelectric Reservoirs (Part 2): Thematic Review," World Commission on Dams, 2000; L. Gagnon, L. and J. Van de Vate, "Greenhouse Gas Emissions from Hydropower," *Energy Policy*, Vol. 25, No. 1, 1997, pp. 7-13.

Conclusions

Pumped hydro is the only electricity storage technology currently deployed on the GW scale worldwide. It has deployment opportunities in the United States measured at least in the tens of gigawatts without limitations of raw materials, limited calendar or cycle life, and with demonstrated AC-AC round-trip efficiencies that routinely exceed 75%. The use of variable speed technology will improve the efficiency and ramping characteristics of new facilities. Closed-cycle plants will reduce environmental impact, and could reduce the permitting time needed. Perhaps the most important R&D effort for PHS is a national screening to assess the total potential for new conventional and unconventional PHS development, especially considering environmental and land use constraints.

Chapter 10: Flywheel Storage

Overview

Flywheels are one of the oldest energy storage technologies, historically used to smooth the power delivery in applications ranging from potter's wheels to reciprocating engines. Flywheel technology has advanced tremendously over the last century, however, bringing the technology from large rotating steel wheels that spin at tens to hundreds of revolutions per minute (RPM) to composite (carbon fiber or fiberglass in resin) rotors that spin up to 100,000 RPM, achieving supersonic speeds in a low pressure vacuum.[359] These advancements allow flywheels to achieve high power and energy densities, high roundtrip efficiency (>80%), low frictional drag losses (<3%/hour), and long operational lifetimes (~20 years) with low operation and maintenance costs.

Flywheels can respond rapidly, both as a source and sink for electricity, making them a valuable resource for frequency regulation in electric power grids. Flywheels are one of the most cost-effective storage technologies for high power (rapid discharge) applications, where they compete directly with batteries. But flywheels are typically more expensive than other storage technologies as an energy resource and are unlikely to compete in the near term for applications requiring several hours of energy storage. Flywheels have also been applied in niche transportation applications, in automobiles, buses, and trains.[360] Flywheels will compete with batteries and capacitors in these markets. Their future market share will depend upon the relative technological advancement in each field.[361]

The rapid charging and discharging characteristics of flywheels, and their ability to cycle hundreds of thousands of times with minimal performance degradation, make them ideally suited to provide power quality and frequency regulation to electric grids. Recent deployment trends have centered around this application. Other applications include distributed uninterruptible power supply (UPS) devices for telecommunication and computing resources, and mobile applications like recycling regenerative braking energy, and using the flywheel to reduce peak power requirements from the motor.[362] Current flywheel research and development are focused on increasing the energy density of flywheels to enable one to several hours of energy discharge, which would enable flywheels to be used in several additional electric utility applications like peak shaving, ramping, and load following.

Beacon Power is the main company developing and deploying flywheels for frequency regulation. **Figure 30** shows Beacon Power's 1MW/250kWh flywheel demonstration project deployed in 2008 to provide frequency regulation to the Independent System Operator (ISO) New

---

[359] Modern flywheel technology is based in part on high-speed gas centrifuge technology developed for uranium enrichment. M.B. Richardson, "Flywheel Energy Storage Systems for Traction Applications," Paper No. 487, International Conference on Power Electronics, Machines and Drives, Bath, UK, April 16-18, 2002.

[360] M.B. Richardson 2002; EPRI/DOE, 2003; M.Lafoz, C. Vazquez, and J. Iglesias, "DC Railway Catenary Regulation Based on KESS," presentation to the Electrical Energy Storage Applications and Technologies Conference, Seattle, WA, October 4-7, 2009.

[361] R. Doucette and M. McCulloch, "A Comparison of High-Speed Flywheels, Batteries, and Ultra-Capacitors on the Bases of Cost and Fuel Economy as the Energy Storage System in a Fuel Cell Based Hybrid Electric Vehicle," *Journal of Power Sources*, No. 196, 2011, pp. 1163-1170.

[362] U.S. Department of Energy, *Flywheel Energy Storage: An Alternative to Batteries for Uninterruptible Power Supply Systems*, DOE/EE-0286, September 2003; M.B. Richardson, 2002; R. Doucette and M. McCulloch 2010.

England electric grid.[363] Beacon expanded this 1MW project to 3 MW in 2009, and also constructed a 20MW/5MWh plant in the New York ISO grid in 2011.[364] Beacon secured an ARRA stimulus grant to develop a second 20 MW plant in the Mid-Atlantic/Midwest grid and proposed additional installations, but filed for bankruptcy protection in October 2011, casting doubt on any of its future projects.[365]

Figure 30. Part of a 1 MW Flywheel System in the ISO-New England Grid

Source: Courtesy of Beacon Power Corp., 2011.

Technology

Description

Flywheels store rotational kinetic energy in the form of a spinning cylinder or disc, then use this stored kinetic energy to regenerate electricity at a later time. The amount of energy stored in a flywheel depends on the dimensions of the flywheel, its mass, and the rate at which it spins. Increasing a flywheel's rotational speed is the most important factor for increasing stored energy—doubling a flywheel's speed quadruples the amount of energy stored.

Flywheels typically use an electric motor to spin a cylindrical rotor at very high speed. Flywheel systems, therefore, consist of the spinning rotor(s), bearings, a motor/generator, power electronics, and a containment enclosure (**Figure 31**). The bearings connecting the rotor to the non-rotating platform are of two types: mechanical bearings, which physically connect the rotor to the housing, or magnetic bearings, which levitate the rotor inside the housing to reduce friction losses. The motor/generator converts electrical energy to rotational kinetic energy to "spin up" the

---

[363] Beacon Power Corp., *Smart Energy Matrix 20 MW Frequency Regulation Plant*, fact sheet, 2010a, http://www.beaconpower.com/files/SEM_20MW_2010.pdf; and Beacon Power Corp., "Beacon Power Corporation," presentation to the Rodman & Renshaw Global Investment Conference, New York, NY, 2010b, http://phx.corporate-ir.net/External.File?item=UGFyZW50SUQ9NDY1Mjd8Q2hpbGRJRD0tMXxUeXBlPTM=&t=1.

[364] The 20MW/5MWh system is composed of two hundred 100kW/25kWh units. Beacon Power Corp., 2010a.

[365] E. Ailworth, "As Federal Funds Dry Up, Beacon Power Still Has Reason for Hope," *Boston Globe*, November 27, 2011.

flywheel, and then regenerates electrical energy at a later time from the spinning rotor. The containment enclosure is the non-rotating platform in which the rotor spins, designed to contain rotor debris in the case of catastrophic rotor failure. The enclosure may also hold a vacuum to reduce air drag on the rotor and standby energy losses.

Figure 31. Electric Flywheel Components

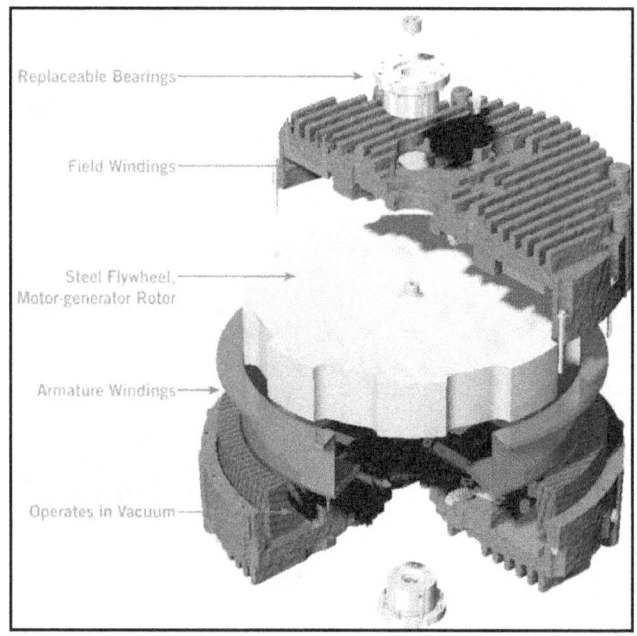

Source: Courtesy of Active Power Corp., 2012.

Performance

Flywheels can be designed for a broad range of applications with different power and energy requirements. These applications can be roughly categorized by their design discharge times. Historically, flywheels were designed for applications with discharge times of about 1 minute, including uninterruptible power supply and transportation applications.[366] More recent applications require discharge times of up to 15-30 minutes for frequency regulation. These applications use composite rotors to achieve high energy densities, along with vacuum housings and magnetic bearings. Long duration flywheels with discharge times of several hours for load shifting have also been considered but have not reached commercial deployment.[367]

Flywheel energy density scales with rotor mass and the square of rotor velocity, based on the physical characteristics of the device. Flywheel power density is largely based on how quickly the rotor can be slowed down. Flywheel roundtrip efficiency is determined by the combination of two factors: instantaneous roundtrip efficiency and standby losses. Instantaneous roundtrip efficiency

---

[366] M.B. Richardson 2002 ; M. Lafoz et al., 2009 ; R. Doucette and M. McCulloch, 2010.

[367] M. Ricci and J. Fiske, *Third Generation Flywheels for Electricity Storage*, DOE award number: DE-FG36-05GO15163, . LaunchPoint Technologies, Inc., Goleta, CA, June 20, 2008; J. Arseneaux, "Development of a 100 kWh/100kW Flywheel Energy Storage Module," presentation to the 2010 Energy Storage Systems Program Conference, U.S. Department of Energy, Washington DC, Nov. 2-4, 2010. http://www.sandia.gov/ess/docs/pr_conferences/2010/agenda.pdf

---

represents the efficiency of turning electricity into rotational kinetic energy, and then returning the kinetic energy back to electricity. Most of the energy losses occur in the power electronics, converting AC electricity to DC electricity, and then the reverse. Typical instantaneous roundtrip efficiencies range from 80% to 90%.[368]

Standby losses refer to frictional losses on the spinning rotor, which increase with the amount of time the rotational energy is stored. Frictional losses are primarily from the bearings and air drag. Typical standby losses range of 2%-3%/hr are primarily managed by using the motor to maintain a constant rotor velocity. For example, to maintain 1 kWh of rotational kinetic energy, a constant electrical load of 25 W may be used to maintain rotor velocity.[369] Employing superconductive magnetic bearings can reduce bearing losses to less than 0.5% per hour, but at higher cost.[370] Because frictional losses transform kinetic energy into heat, and the constant electrical load required to maintain a constant rotor speed transforms to heat, device temperature must be managed to keep the rotor within sustainable limits.

Since standby losses are dependent on the amount of time energy is stored, they vary significantly depending on application. For example, a flywheel dispatched in frequency regulation markets can have approximately 6,000 full charge/discharge cycles in one year with a mean stored energy time on the order of 1.5 hours.[371] In this application, roundtrip efficiency would likely be in the range of 80% (85% instantaneous minus 5% standby losses) and very competitive with other technologies, namely batteries, operating in this market. If flywheels were used to shift off-peak generation to meet on-peak load, flywheel roundtrip efficiencies could decrease to 50%-70% (85% instantaneous minus 15% for five hours of storage or minus 36% for twelve hours of energy storage), reducing their economic feasibility.

Flywheels can have very rapid response times; those designed for frequency regulation can ramp to full nameplate power capacity in less than four seconds.[372] Because flywheels can ramp more rapidly than conventional generators, a smaller amount of flywheel power capacity can provide the same ramping characteristics for regulation as a larger amount of conventional generation capacity.[373]

Flywheels have minimal operations and maintenance costs, design lifetimes on the order of 20 to 25 years, and are designed to be cycled hundreds of thousands of times with little or no performance degradation.[374] Flywheels also have high operational reliability. In a recent demonstration project, frequency regulation flywheels were available for more than 98% of the

---

[368] EPRI/DOE, 2003; C. Lyons, "Flywheel Energy Storage—A Smart Grid Approach to Supporting Wind Integration," presentation to the Electrical Energy Storage Applications and Technologies Conference, Seattle, WA, October 4-7, 2009.

[369] KEMA, Inc., *Emissions Comparison for a 20 MW Flywheel-based Frequency Regulation Power Plant*, 2007a.

[370] M. Strasik, et al., "Design, Fabrication, and Test of a 5 kWh/100 kW Flywheel Energy Storage Utilizing a High-Temperature Superconducting Bearing," *IEEE Transactions on Applied Superconductivity*, Vol. 17, No. 2, 2007, pp. 2133-2137.

[371] Beacon Power, Inc., 2010b.

[372] J. Eyer, *Benefits from Flywheel Energy Storage for Area Regulation in California-Demonstration Results*, SAND2009-6457, Sandia National Laboratories, 2009.

[373] Y.V Makarov, et al., "Assessing the Value of Regulation Resources Based on Their Time Response Characteristics," PNNL – 17632, Pacific Northwest National Laboratory, June 2008.

[374] Beacon Power Corp., 2010b; J. Eyer, 2009.

time.[375] The most common flywheel failure mode is the propagation of cracks through the rotor over time. The presence of cracks can frequently be diagnosed by reduced performance in composite rotors before hazardous failure occurs.

Table 11 summarizes flywheel performance characteristics. Flywheel sizes, shown by flywheel energy and power footprints, vary significantly depending on the energy and power characteristics of each application. Roundtrip efficiencies also vary significantly across applications because flywheels for applications like load shifting would typically designed to reduce standby losses (at the additional cost of high speed composite rotors, and magnetic bearings) whereas frequency regulation flywheels are designed with higher standby losses. Load-shifting and frequency regulation flywheels also are dispatched differently. Flywheels generally have similar design lifetimes and operations and maintenance (O&M) costs across a range of sizes and applications.

Table 11. Flywheel Performance Characteristics

Characteristic	Value	Comments
Power footprint (kW/m²)	1.4 – 490	Lower value for frequency regulation with 15 minute discharge. Upper value for a power application with 1 second discharge.
Energy footprint (kWh/m²)	0.35 – 0.54	Lower value for frequency regulation with 15 minute discharge. Upper value for a power application with 1 second discharge.
Roundtrip efficiency	80%-90%	Instantaneous roundtrip efficiency for frequency regulation flywheels. Roundtrip efficiencies vary by application because of the different standby losses associated with different charge and discharge characteristics.
Standby energy loss	1%- 3%/hour	Standby losses represent frictional losses in an extended discharge flywheel, where it frequently takes 10-30W of continuous power to maintain 1 kWh of stored energy. Hourly standby losses are typically higher in rapid discharge flywheels.
Ramp rate	>25% / sec	Frequency regulation application.
Operational lifetime	>100,000 cycles, >20 years	Rapid discharge and extended discharge flywheels have similar design lifetimes.

Sources: Beacon Power Corp., 2010a; J. Eyer, 2009; EPRI/DOE, 2003; NREL analysis.

Cost

Flywheel costs can vary significantly depending on the power and energy requirements of each application. While flywheels are an old technology, they have only recently been applied to electricity storage. Current costs are largely based on experience building demonstration projects. Significant near-term cost reductions are projected due to organizational learning and economies of scale in future deployments. An example of current costs is the 20MW/5MWh demonstration project in the New York ISO, with a cost of $69 million (2009 dollars). Beacon Power has set a cost target of $25 million-$30 million for an "n[th] system" frequency regulation plant, as shown in **Figure 32.** The plant is designed with 15 minutes of discharge capacity, corresponding to cost projections, on an energy basis, of $14,000/kWh for the first demonstration plant to $5,000-

---

[375] Beacon Power Corp., 2010b.

$6,000/kWh for the nth plant. Long-term O&M costs have yet to be determined. Estimates include $10-$12/kW-yr for fixed O&M and about $3/MWh of output for variable O&M.[376]

Figure 32. Flywheel Cost Projections for a 20MW/5MWh Frequency Regulation Plant

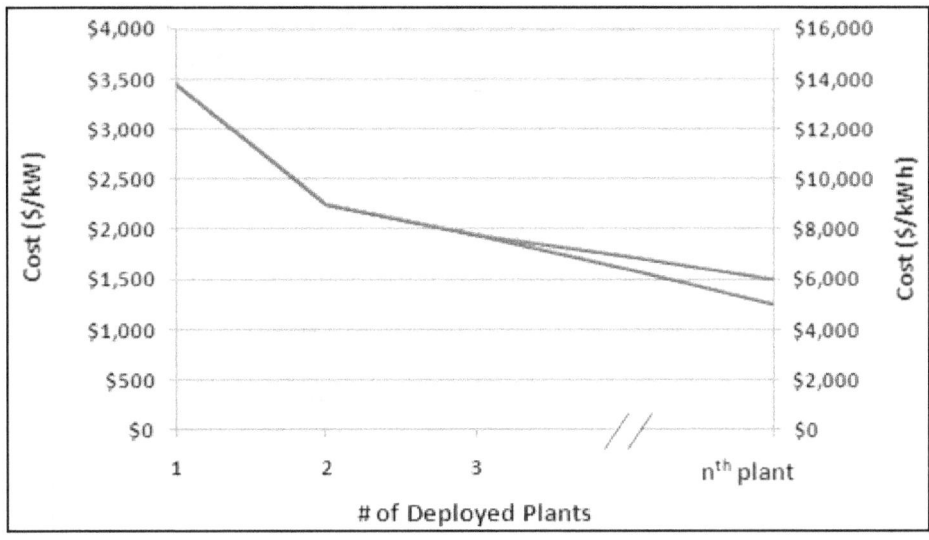

Source: National Renewable Energy Laboratory analysis, 2011; Beacon Power Corp., 2010b.

Research and Development

Recent flywheel research and development (R&D) have focused on improving flywheel energy density, reducing standby losses, increasing the efficiency of associated power electronics, and decreasing cost.

To improve flywheel energy density, R&D has focused on developing and using new rotor materials capable of sustaining higher speeds. Energy density has improved significantly by transitioning from metallic rotors, with maximum rim velocities of 300-500 m/s, to composite rotors which can achieve rim velocities well above 1,000 m/s. **Table 12** summarizes flywheel rotor material characteristics, including maximum theoretical energy densities, and the actual or proposed energy densities for real flywheel systems. Operating flywheels have lower energy densities than the maximum theoretical limits because flywheels are operated below their maximum stress threshold for safety reasons (typically a reduction of 25%-50% in flywheel performance).[377] Furthermore, realistic rotor designs are not fully able to utilize a material at its stress potential at each position within the rotor (typically an additional 20%-50% performance reduction).[378] This leads to current flywheels performing at 25% to 60% of the maximum theoretical limits. The difference between real world performance and the maximum theoretical

---

[376] EPRI/DOE, 2003, Eyer 2009

[377] P. Johnson, et al., "Design, Fabrication, and Test of a 5 kWh Flywheel Energy Storage System Utilizing a High Temperature Superconductive Magnetic Bearing," proceedings of the Electric Energy Storage Applications and Technologies Conference, Sandia National Laboratories, San Francisco, CA, October 17-19, 2005.

[378] F. Werfel, et al., "Towards High-Capacity HTS Flywheel Systems," *IEEE Transactions on Applied Superconductivity*, No. 20, 2010b, pp. 2272-2275.

---

performance roughly represents the R&D potential for improved design, although flywheels will always be operated below their maximum stress thresholds for safety considerations, and there are other fundamental limits to design improvements.

Table 12. Flywheel Materials Characteristics

	Maximum Theoretical Energy Densities (Wh/kg)	Achieved or Proposed Energy Densities (Wh/kg)
Carbon Nanotube	—	2,700
Fused Silica	—	800
Graphite	373-532, 547-780	230
Spectra 1000	430	
Kevlar 49	351-370	
Fiberglass	160-225	40
Steel (4340)	32-47	30

Sources: M. Strasik et al. 2007; J. Bitterly, "Flywheel Technology: Past, Present, and 21st Century Projections," *IEEE Aerospace and Electronic Systems Magazine*, August, 1998; H. Liu and J. Jiang, "Flywheel Energy Storage – An Upswing Technology for Energy Stability," *Energy and Buildings*, No. 39, 2007, pp. 599-604; F. Werfel et al., "HTS Flywheel from R&D to Pilot Energy Storage System," *Journal of Physics, Conference Series*, Vol. 234, 2010a.

Note: Energy density is shown here in Watt-hours (Wh) per kilogram (kg) of rotor material. Flywheels designed for vehicle applications often measure energy density in Wh/kg, while larger flywheel plants for frequency regulation may measure energy density in units of Wh/m^2. Wh/kg provides the more useful metric to compare the energy density of rotor materials

Another area of active R&D, not limited to the flywheel community, is developing high tensile strength materials. For example, carbon nanotubes can reach theoretical tensile strengths that are an order of magnitude higher than graphite.[379] While these materials would have to be demonstrated, and cost-effectively manufactured at scale, they could considerably increase flywheel energy density.

In addition to improving rotor materials, new rotor designs are being developed to increase energy density. Design trends include the use of larger diameter rotors to increase rim tip speeds, and eliminating the central shaft and hub, levitating the rotor on magnetic bearings located at or near the rim to better utilize the tensile strength of rotor materials.[380] This approach could provide additional controllability for large diameter energy flywheels and could reduce several engineering challenges, including rotor rigidity and exciting frequency oscillation modes.[381] While such flywheels have not been demonstrated, it is projected that new designs could significantly improve energy density and flywheel costs. Flywheel rotor R&D is being conducted in the United States by universities (e.g., the University of Texas and Pennsylvania State University), Sandia National Laboratories, and the U.S. Army, as well as several commercial companies including Boeing, Beacon Power, and LaunchPoint, Inc.[382]

---

[379] J.P. Salvetat, et al., "Mechanical Properties of Carbon Nanotubes," *Applied Physics A.*, No. 69, 1999, pp. 255-260.

[380] M. Ricci and J. Fiske, 2008; J. Arseneaux, 2010.

[381] M. Ricci and J. Fiske, 2008.

[382] EPRI/DOE, 2003; T. Boyle et al., "Improved Properties of Nanocomposites for Flywheel Applications," presentation to the Energy Storage Systems Program Conference, U.S. Department of Energy, Washington DC, Nov. 2- (continued...)

To reduce standby energy losses, flywheel R&D has focused on reducing frictional drag in the bearings and air drag on the spinning rotor. Air drag can be effectively reduced by pulling a vacuum in the rotor housing—the approach used in current designs. Bearing friction can be reduced by replacing mechanical bearings with magnetic bearings that levitate the rotor inside its housing.[383] Magnetic bearings typically consist of a fixed magnet on the rotor, and a fixed and/or electromagnet on the rotor housing. Magnetic bearings are either static (passive system with a fixed magnetic field that holds the rotor in place, with additional control components), active (variable magnetic field that dynamically controls flywheel position and motion), and/or superconducting (passive or active electromagnet designed using high temperature superconductive components).[384] These bearings either draw constant current (active magnetic bearings) to offset resistive losses, or can be designed using high-temperature superconductive materials which draw current to power cryogenic coolers.[385] Magnetic bearings can reduce standby losses to a few percent per hour where losses represent the combination of the remaining frictional losses and the system load required to operate magnetic bearings.[386] Superconducting bearings can reduce standby losses down to 0.1%/hour.[387] Flywheel bearing R&D continues at the Lawrence Livermore National Laboratory as well as international universities and several commercial companies.[388]

Flywheel power electronics have become more efficient over time. Recent power electronics R&D has been driven by other, larger industries, and flywheels benefit from the resulting incremental efficiency and durability improvements. These improvements will be shared with other storage technologies that do not use conventional generators.

Future flywheel R&D will likely focus on further improving flywheel energy density, magnetic bearing technology and control circuits, and reducing materials and manufacturing costs. It is likely that flywheels will see incremental efficiency improvements, driven by efficiency gains in the power electronics and bearings. Flywheel energy density could see larger improvements, particularly by using new materials capable of higher energy densities and by modifying flywheel design for extended discharge applications.[389] In addition to fundamental R&D improvements, there is a large potential for reducing costs based on organizational learning.

(...continued)

4, 2010; J. Tzeng,, R. Emerson, and P. Moy, "Composite Flywheels for Energy Storage," *Composites Science and Technology*, No. 66, 2006, pp. 2520-2527; M. Strasik, et al., 2007b; J. Hull et al., "High Rotational-Rate Rotor with High-Temperature Superconducting Bearings," *IEEE Transactions on Applied Superconductivity*, No. 19, 2009, pp. 2078-2082; M. Ricci and J. Fiske, 2008.

[383] Use of magnetic bearings also eliminates lubrication challenges. As flywheel spin rates exceeds about 40,000 RPM, the lubrication in mechanical bearings (mostly ball bearings) begins to heat and fail. H. Liu and J. Jiang, 2007.

[384] Electric Power Research Institute (EPRI), *Flywheels for Electric Utility Energy Storage*, TR-108889, Palo Alto, CA, 1999.

[385] J. Hull et al., 2009.

[386] Beacon Power Corp., 2010b.

[387] P. Johnson et al., 2005.

[388] Lawrence Livermore National Laboratory, "LLNL and Arnold Magnetic Technologies Collaborate on Passive Magnetic Bearing System," press release, May 26, 2010; N. Koshizuka, "R&D of Superconducting Bearing Technologies for Flywheel Energy Storage Systems," Physica C, No. 445, 2006, pp. 1103-1108.

[389] J.P.Salvetat et al., 1999; T. Boyle, et al., 2010; M. Ricci and J. Fiske, 2008; J. Arseneaux, 2010.

Deployment Challenges

Flywheel permitting should be relatively easy compared to several other technologies since flywheels do not use fuel or contain hazardous chemicals, do not use water, are emissions free, and have a relatively small footprint.[390] Flywheel units up to 10 kWh can be assembled in a factory and delivered using standard shipping trucks.[391] Both the modular nature of flywheels, and low operations and maintenance requirements, allow flywheels to be sited virtually anywhere. One estimate of complete siting, permitting, and construction time is about 18 months.[392]

The physical footprint of frequency regulation flywheels is about 50 kW/m^2 and or about 10 kWh/m^2. The footprint of the entire regulation plant (including power conditioning equipment and a security perimeter) is about 6 MW/acre (for example, a recent 20 MW/5MWh regulation flywheel plant is sited on about three acres).[393] The relatively small land area requirement (and their modular nature) allows flywheels to be located at substations, or near urban load centers. In act, flywheels are frequently sited underground encased in thick concrete walls to contain debris should a catastrophic failure occur.[394] While steel flywheels are subject to hazardous failure modes, composite flywheels have a diagnosable deterioration in performance before failure occurs.[395] With these precautions, flywheels can be operated safely and reliably through their 20+ year calendar life. Flywheels do not contain any toxic materials, and there is no risk for explosion or environmental contamination.

Flywheels are primarily constructed from metal, carbon, or glass fiber and resin, and support materials like concrete and structural steel. While these materials will be subject to commodity prices, their supply is virtually unlimited. High temperature superconductive materials like yttrium barium copper oxide (YBCO) use common earth elements, and material shortages are not a concern. Neodymium magnets are used in some motors and magnetic bearings, and may have a limited supply. However, different magnetic materials or electromagnets can also be used.

Flywheel systems are being deployed now using a basic utility engineering and construction labor force. While flywheel deployment will lead to new jobs, it will not require unique new infrastructures or labor forces.

Conclusions

Several demonstration projects are currently showing that flywheels are a competitive source of frequency regulation. The rapid response of flywheels, and their ability to cycle extensively with

---

[390] KEMA, Inc. 2007a.

[391] J. Hull, *Flywheels, Encyclopedia of Energy, Volume 2,* Elsevier, Inc., ISBN:978-0-12-176480-7, 2004.

[392] C. Lyons, "A Smart Grid Approach to Regulation and Ramping," *Renewgrid*, August 2009.

[393] Beacon Power Corp., 2010b.

[394] Flywheels store a very large amount of energy, which could be destructive in a catastrophic failure. For example, a 100 kW Smart Energy 25 flywheel from Beacon Power stores 25 kWh of rotational kinetic energy when fully charged C. Lyons, 2009. This amount of energy is approximately equal to 20 kg of trinitrotoluene (TNT). Flywheel housing and encapsulation material must, therefore, be designed to contain debris should catastrophic failure occur.

[395] Composite rotors are designed so that stress fractures propagate longitudinally through the material, leading to the delimitation of outer rotor layers prior to failure, causing the system to wobble and slow down through frictional losses before catastrophic failure occurs.

little or no performance degradation, makes them ideally suited to provide regulation. However, they will compete with batteries and supercapacitors in these markets, and flywheel economics will have to be (and remain) dominant over these technologies to capture and maintain market share. Other potential markets for flywheel energy storage include transportation (capturing regenerative braking energy), and additional ancillary services that bridge regulation and reserves markets and require a longer-duration discharge.

Flywheel R&D will likely further improve energy density, reduce standby energy losses, and reduce costs. Energy density improvements will be driven by the development and use of new composite rotor materials, and larger diameter rotors. Standby losses will likely be reduced using superconducting magnetic bearings. Flywheel costs will likely be reduced by implementing new rotor materials and designs, and from organizational learning as more flywheels are deployed and the industry benefits from economies of scale.

Chapter 11: Thermal Energy Storage in Buildings

Overview

Thermal energy storage (TES) in buildings is an alternative to most electricity storage technologies which is often overlooked because it does not store and discharge electricity directly. However, in some applications, TES can be functionally equivalent to electricity storage, with efficiencies that exceed 90%. One key application is to use cold storage to shape end-use electrical demands in the same manner as customer-sited electricity storage, with corresponding benefits of reduced distribution capacity and losses, and with higher efficiency and potentially lower cost.

Air conditioning for buildings represents a significant component of end-use of energy in the United Sates, comprising approximately 10% of total electricity sales.[396] More importantly, the peak demand for most U.S. utilities is strongly driven by the aggregated cooling demands of individual buildings. "Cool" thermal storage relies on a simple concept: operate a building's energy-intensive refrigeration equipment for space conditioning at night to create and store energy in the form of ice or chilled water. That chilled water is then used as a reservoir of "cold" during the day to meet building cooling loads, reducing the operation of the building's cooling equipment during daytime on-peak periods. In effect, the TES shifts electricity demand from day to night, when electric energy cost and utility demand is low, in a highly predictable way. The concept of cool storage as a "thermal battery" translates into operating cost savings for the building owner. It also yields broader benefits for utilities when widely applied through its ability to decrease the aggregate electric demand during on-peak periods. By decoupling the production of cooling with the demand for cooling, TES offers the opportunity to considerably alter the electrical demand profile associated with building space conditioning systems. Heat storage technology similarly offers potential benefit for load shaping of buildings that derive heating energy from electricity (either by resistance heating or heat pumps). This application of thermal energy storage is not mature or widely applied but does offer potential—particularly for winter-peaking electric utilities.

In the United States, there are approximately a dozen manufacturers or providers of thermal energy storage systems (ice storage and chilled water storage). The industry trade association currently has five U.S. thermal energy storage manufacturers as members of their "Thermal Storage Equipment" product section.[397] Most of the TES technologies being sold into the U.S. market today are mature, having undergone refinements over the past three decades. One notable exception are unitary ice thermal storage systems, which integrate the required refrigeration system along with a thermal energy storage module. Unitary systems have been commercialized only relatively recently.

Although detailed statistics about the cool thermal energy storage market in the United States are not available, an estimated 4,500 thermal energy storage systems are operating across the country. There are also a number of demonstrated facilities in federal and public buildings.[398] The primary

---

[396] Energy Information Administration, *Annual Energy Outlook 2011*, DOE/EIA-0383(2011), 2011.

[397] The Air-Conditioning, Heating, and Refrigeration Institute (AHRI) is a trade association representing more than 300 manufacturers of air conditioning, heating and commercial refrigeration equipment.

[398] W. D. Chvala, *Technology Potential of Thermal Energy Storage (TES) Systems in Federal Facilities*, PNNL-13489, (continued...)

---

challenge of TES is due to its distributed nature—deployment is dependent on individual customers being aware of its existence and current electricity rate structures typically do not capture its benefits to the grid. Partly in recognition of this challenge, in 2010 a consortium of municipal utilities in southern California began installing 53 MW of distributed ice storage systems as a system resource for load shifting and firm capacity.[399] This consortium represents a potential shift in approaches to distributed storage, as it may be easier to develop business models that deploy utility-owned but customer-sited storage.

Technology

Description

Broadly speaking, TES technologies are classified into two categories based on the nature of how they store energy: "sensible" energy change and "latent" energy change. Sensible energy change systems store thermal energy by using the heat capacity of a working fluid (e.g., water) and causing it to undergo a temperature change. During charging, warm water from the top of a storage tank is cooled using a mechanical refrigeration plant (chiller) prior to being returned back to the bottom of the tank. The cool water within the water storage tank is then made available to meet instantaneous building cooling loads at a later time. As the water absorbs heat when meeting the building's space cooling load, its temperature rises and the warm water is returned back to the top of the storage device for re-cooling during the off-peak period. **Figure 33** shows a simple schematic of a chilled water storage system operating to meet building cooling loads.

Latent energy change technologies extract and absorb heat into a storage medium that undergoes a liquid-solid phase change (e.g., melting and freezing). The most widely used latent energy change storage medium is ice but other substances, like salt solutions or ethylene glycol-water mixes, have been employed as well. During charging, the chiller cools the fluid—say, an ethylene glycol-water solution—to a temperature below the freezing point of water. The cold glycol circulates through a heat exchanger immersed in a water tank causing it to freeze. During a melt period, warm glycol returning from the load can either be pre-chilled by the mechanical chiller or cooled directly by circulation through the ice storage tank(s). The warm glycol is cooled as it gives up its heat to the ice causing it to melt back to a liquid.

(...continued)

Pacific Northwest National Laboratory, July 2001; U.S. Department of Energy, "Thermal Energy Storage at a Federal Facility," DOE/GO-102000-1099, fact sheet., July 2000; Stanford University, "Ice Plant," web page, 2011. http://lbre.stanford.edu/sem/ice

[399] Ice Energy, "SCPPA to Undertake Industry's Largest Utility-Scale Distributed Energy Storage Project," press release, January 27, 2010. http://www.ice-energy.com/content10197

Figure 33. Illustration of a Chilled Water-Based TES System

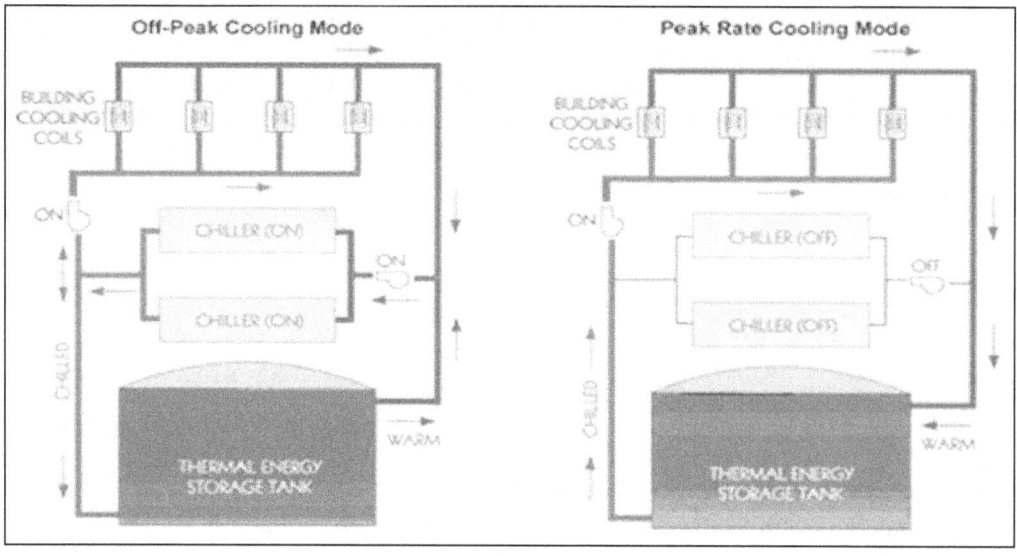

Source: Courtesy of Cypress, Ltd., "Stratified Chilled Water Storage," web page, Coto de Cazxa, CA, 2011, http://www.shiftnsave.com/pge/schws.php.

Heat storage is an even simpler form of TES. It stores heat in a high-heat capacity material, such as bricks, and discharges that stored heat later.[400] While much simpler than ice storage, it has several disadvantages. First, while virtually all air-conditioning is driven by electricity, a large fraction of space heating is accomplished by burning natural gas or other fuels, reducing the potential opportunities. Second, hot storage has fewer utility benefits, especially in terms of providing system capacity. Nearly all of the United States is summer peaking, meaning there is generally large capacity reserves to meet winter demand. However, hot storage can still provide valuable load-leveling applications. Controllable water heating is another form of TES, which somewhat blurs the line between energy storage and demand response.

Performance

Historically, thermal storage has been used to reduce demand and peak energy charges, making it functionally equivalent to an electricity storage device providing load-leveling and firm capacity. However, while charging, TES has the potential to provide operating reserves as well. The compressor for cold storage, for example, could respond to a signal for frequency regulation or spinning reserves while charging and reduce load or turn off. Given the flexibility of when the device charges during the overnight period, this should have limited effect on the customer.

Figure 34 shows how two alternative cool storage system operating strategies can alter the electrical demand for a facility compared to a building without thermal energy storage. Without TES, the peak demand associated with its cooling system operation is 500 kW with the peak occurring mid-afternoon. If TES was added to the building, it would allow the refrigeration plant to idle during the on-peak period yielding a 500 kW demand reduction (full-storage). A somewhat smaller TES system could be installed that would necessitate some chiller operation during the on-peak period but still yield a 50% reduction in electrical demand during the time of the day

---

[400] For an example in residential applications see Steffes Corp., "How IT Works: ETS Heating System," web page, 2011. http://www.steffes.com/off-peak-heating/how-ets-off-peak-heating-works.html

when utilities experience their peak demand (partial storage). The size of a TES influences the extent to which it can alter a building's demand profile. This gives designers flexibility to achieve end-use objectives that may vary from building to building or utility region to utility region.[401]

Figure 34. Electric Demand for Building Cooling With and Without TES

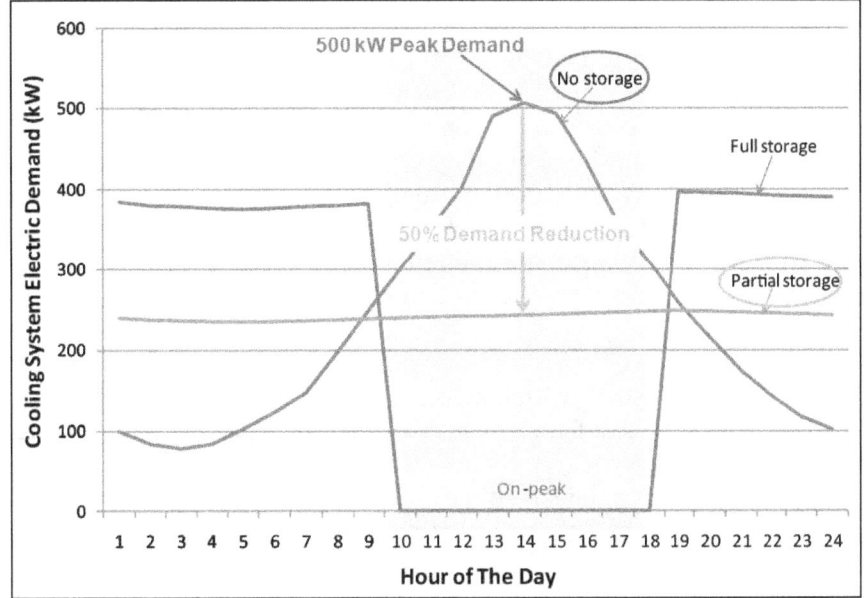

Source: D. Reindl, University of Wisconsin-Madison, 2011.

In all performance metrics, TES must be compared to the system it is replacing. The concept of "round trip efficiency" is not easily defined in TES systems in part because it must be compared to the conventional alternative. However, by many measures the efficiency of TES is commonly cited as well above 90%. The overall energy "effectiveness" can sometimes even be higher than 100% because a cold-storage based cooling system can use less electricity than its conventional alternative. Unlike most technologies, TES does not have to convert electricity into an intermediate energy carrier and back to electricity. TES converts electricity into "cold" in essentially the same manner it would in a system without TES, only at a different time. As a result, the primary losses (again compared to its conventional alternative) are heat losses in the storage tank (which are generally well insulated). One estimate of heat losses in a commercial ice storage tank was about 0.7% per day.[402] Another report claims 1%-5% per day.[403] However, TES systems can improve the operating efficiency of equipment used for air conditioning compared to non-storage systems. A building with TES system will run its refrigeration equipment at nighttime when the system efficiency is higher due to cooler ambient conditions.[404] In addition TES system

---

[401] A. Wilson, "Buildings on Ice," *Environmental Building News*, Vol. 18, No. 7, July 2009, http://www.calmac.com/documents/MakingTheCaseForThermalEnergyStorage_EnvironmentalBuildingNews.pdf.

[402] T.K. Stovall, *Calmac Ice Storage Test Report*, ORNL/TM-11582, Oak Ridge National Laboratory, 1991.

[403] K. Roth, R. Zogg, and J. Brodrick, "Cool Thermal Energy Storage," *ASHRAE Journal*, September 2006.

[404] This is due to the fact that "cold" is actually made by removing heat from the cooling medium and discharging that heat into the outside environment. It is easier and more efficient to discharge the removed heat into cooler air, and as a result a device making cold in the cooler evening is more efficient than making cold in the heat of the day. See R. Willis and B. Parsonnet, "Energy Efficient TES Designs for Commercial DX Systems," *ASHRAE Transactions*, OR-10-016, June 8, 2010; M. MacCracken, "Thermal Energy Storage Myths," *ASHRAE Journal*, September 2003.

allows operation of refrigeration equipment at or near peak efficiency during all operating hours compared to non-storage systems which have to operate at less efficient part-load conditions for the majority of its hours of operation annually.[405] The net effect is the potential to decrease total energy consumption associated with cooling, in addition to the load-shifting and capacity benefits.

There are examples of comparable buildings where the TES equipped building uses less cooling energy than the non-TES building, implying an efficiency of greater than 100%. One case involves the installation of an ice storage TES system at one of the Kraft Foods Headquarters buildings in Northfield, IL. Two comparably constructed buildings were operated—one with ice TES and one without. In 1997, the building equipped with ice storage consumed 10,114,460 kWh with a peak demand of 2,368 kW while the non-storage building required 11,695,468 kWh (+15%) with a peak electrical demand of 3,307 kW (+40%). The TES equipped building operated during that year with a 19% ($193,000) lower electric bill.[406] In addition, TES has the added benefit of reduced transmission and distribution losses similar to other distributed energy storage devices.[407]

For most TES technologies, the device lifetime is equal to or longer than the refrigeration system it is displacing. Detailed O&M cost figures (which again must be compared to the O&M requirements to that alternative system) are not widely published, however, the ongoing maintenance of a TES system is minimal—often only requiring occasional checks of water levels and water quality to assure adequate treatment is being sustained. Furthermore, buildings equipped with TES often have fewer chillers or smaller chillers. This means that there is less "rotating mechanical equipment" to maintain, so O&M for a TES equipped system can be less than a non-TES system.

Cost

As with other performance metrics, the cost of a cooling system with TES must be compared to the system without one. The cost associated with addition of a TES is partially offset by reductions in the size (or number) of active refrigeration or chilling equipment that would otherwise be required to meet an application's peak load directly. In fact, a number of building thermal storage systems have been installed at a lower capital cost when compared to their non-storage system counterparts.[408] The actual costs of storage systems will be variable depending on the sizing strategy. One estimate for the installed cost for ice storage systems ranges from $500 to $1,000 per kW of electricity shifted from on-peak to off-peak periods kW.[409] This cost range does not include the emerging unitary ice storage systems with recent estimates in the range of $2000-$3000/kW.[410] Chilled water storage system costs range from $330-$1,350 per kW.[411]

---

[405] "ASHRAE GreenGuide," American Society of Heating, Refrigerating, and Air-Conditioning Engineers, Atlanta, GA, 2003, pp. 84-85.

[406] İ. Dinçer and M. Rosen, *Thermal Energy Storage: Systems and Applications*, John Wiley & Sons, 2002.

[407] Gansler, R. A., Reindl, D. T., and Jekel, T. B., "Simulation of Source Energy Utilization and Emissions for HVAC Systems", ASHRAE Transactions, V. 107, Pt. 1, 2001.

[408] J.S. Andrepont, "Stratified Low-Temperature Fluid Thermal Energy Storage (TES) in a Major Convention District—Aging Gracefully, as Fine Wine," *ASHRAE Transactions* Vol. 112, No. 1, 2006.

[409] Electric Power Research Institute (EPRI), "Thermal Energy Storage Technology Brief," No. 1016084, Palo Alto, CA, November 2008.

[410] M. Wald "Storing Energy as Ice?," *Green* (internet blog), *New York Times*, January 27, 2010; Ice Energy, "Glendale (continued...)

Research and Development

The state of cool storage technologies today can be described as "semi-mature." A considerable amount of technology development occurred in the 1980s with subsequent refinements that have continued to date. The technologies currently available are highly functional and further development efforts are not critical to the future success of the technology. However, there are several R&D efforts that could provide incremental improvement in TES performance including:

- **Advanced heat transfer fluids.** Ice storage systems principally rely on the use of ethylene glycol or propylene glycol heat transfer fluids. Thermal storage systems and the broader industry could benefit appreciably by the development of high heat carrying capacity secondary fluids that are non-toxic with low viscosity. This includes exploring the potential for using nanoparticles or micro-encapsulated phase change materials to increase the fluid's heat capacity.

- **Efficient low temperature refrigeration.** By their nature, ice storage systems require lower operating temperatures to charge the storage system. Advances aimed at increasing the operating efficiency of refrigeration technologies at lower operating temperatures can further improve the energy advantage of ice storage TES technologies compared to non-storage system counterparts. This area has not received attention in the industry because of the focus on higher operating temperature systems used for direct space conditioning.

- **Communication and Control.** Like other distributed storage technologies, TES will provide maximum benefit when responding to the needs of the grid as a whole, as well as the building cooling demand. There are opportunities to optimize the aggregated charging and discharging of thermal energy storage systems to maximize grid stability and generation dispatch via intelligent communication (i.e., smart grid).

Deployment Challenges

The primary barrier to TES deployment is associated with the fact that it is customer sited and deployed. There is currently a lack of technology awareness among end users, as well as policy makers and utilities that might encourage TES adoption. For example in several of the major surveys of energy storage, TES is not even listed as an option.[412]

In the late 1980s and early 1990s, EPRI and its member utilities conducted a significant amount of outreach and education through the Thermal Storage Applications Research Center (TSARC) to a wide range of stakeholders. These outreach efforts drove a significant number and variety of applications. By the mid 1990s, outreach support for thermal storage ended and visibility of

---

(...continued)

Water & Power to Launch Thermal Energy Storage Project," press release, March 16, 2010. http://www.ice-energy.com/content10195

[411] EPRI, 2008.

[412] Several of the common storage resources including the EPRI-DOE Handbook, the Electricity Storage Association website, and recent storage reports by the Electricity Advisory Council, and the American Physical Society do not even mention thermal storage. The historical lack of awareness of TES as an option is noted in: C. E. Dorgan, R.T. Linton, and S.L Mattix, *Market Assessment of Thermal Energy Storage Report*, No: TSARC 96-01, Thermal Storage Applications Research Center, University of Wisconsin-Madison, May 1996.

thermal storage technology, its benefits, and sustained application declined significantly. This history underscores the need to develop and support effective outreach efforts to encourage application of the technology.

For TES to be incorporated effectively, building designers must be aware of and willing to consider the technology. Currently, building design tools are not widely available to support quick and accurate systems analysis. Many of the most widely used software tools for building mechanical systems do not include TES technologies. Designers must also be willing to address design complexity. Although the added complexity of TES to a building is small to moderate, many designers perceive this added complexity as increased risk. Overall, greater awareness and suitable design tools are needed to increase consideration of TES as a design option. The addition of a cool storage system necessitates a greater footprint within or outside of a building compared to a non-storage system alternative. This reality can be overcome with proper planning and coordination with architects and building designers—even in cases where space is constrained.

Another major challenge of TES is the limited quantification and recovery of benefits. Thermal storage systems become cost-justified in cases where demand costs are high and time-of-use rates exhibit high energy cost differentials between on-peak and off-peak periods. When rate structures do not completely capture the capacity and time-shifting value of TES, then the technology will be undervalued and adoption minimized.[413] For building types that would utilize small TES technologies, time-of-use rate structures are not uniformly available; therefore, this technology is particularly challenged to achieve market penetration. As a result, one manufacturer has developed a business model in which customer-sited storage is owned by the utility as a peak generation and load shifting asset.[414]

There are few other non-technical barriers to large-scale deployment. TES requires little in the way of exotic materials, and the only potentially hazardous materials are certain refrigerants, which are common to the industry. There are no discharges of any material into the environment for a well maintained system, as with other cooling technologies. There is also the associated (small) risk of tank failure and discharge of the chilled water.

Conclusions

As a technology, TES has several advantages over conventional electricity storage devices. Most importantly, it effectively stores energy at much higher round trip efficiencies and can also be deployed at the point of use, decreasing need for transmission and lowering transmission and distribution losses. The primarily disadvantage of thermal storage is that it is tied to an end use (air-conditioning), which means it is less flexible than other electricity storage technologies which can provide storage services that are unrelated to the demand for cooling or heating. The primary barriers to deployment are not technical or economic—given appropriate valuation, benefits of TES can outweigh the costs with existing technology. However, limited awareness, technology perception, and limited access to recovery of benefits currently restrict adoption and suggest market and policy changes are needed to allow equitable competition for TES.

---

[413] For a more comprehensive discussion of the benefits of customer-sited TES see J. Swisher, W. Clift, and E. Lokey, *System Benefits of Mass Deployment of Distributed Thermal Energy Storage: An Evaluation of the Physical and Environmental Impacts of the Ice Power Plant*, Camco International, June, 2009; and R.W. Beck, *Ice Bear Energy Storage System Electric Utility Modeling Guide*, October 8, 2010.

[414] Ice Energy, January 27, 2010.

Chapter 12: Thermal Energy Storage for Concentrating Solar Power

Overview

Another application of thermal energy storage for electric grid applications is storing thermal energy from the sun that is later converted into electricity. Incorporating thermal energy storage (TES) into concentrating solar power (CSP) plants enables these plants to dispatch power beyond their normal operational hours of daytime sun. Such a system can be functionally equivalent to other energy storage technologies for the grid. TES allows full-load generation hours to be added or shifted, providing increased utilization (capacity factor) of the power block, providing firm capacity, generating higher-value electricity, and potentially reducing the levelized cost of energy.[415]

CSP with TES has been commercially deployed in three Andasol parabolic trough solar plants in Spain (**Figure 35**).[416] Each of the three Andasol plants is rated at a nominal power output of 50 megawatts (MW) of electricity with storage that provides an additional generating capacity of 7.5 hours.[417] The Gemasolar plant, completed in Spain in 2011, is a 17 MW power tower with 15 hours of TES; it is the first commercial power tower with molten-salt heat-transfer fluid (HTF) and storage.[418]

---

[415] R. Sioshansi and P. Denholm, "The Value of Concentrating Solar Power and Thermal Energy Storage," *IEEE Transactions on Sustainable Energy*, Vol. 1, No. 3, pp. 173-183, 2010.

[416] M. Medrano, et al., "State of the Art on High Temperature Thermal Energy Storage for Power Generation. Part 2—Case Studies," *Renewable and Sustainable Energy Review*, No. 14, pp. 56-72, 2010.

[417] As a result of this single plant, the capacity of CSP TES now exceeds most battery technologies.

[418] NREL 2011, "Concentrating Solar Power Projects." http://www.nrel.gov/csp/solarpaces/

Figure 35. Two-Tank TES System at a 50 MW Solar Power Plant in Spain

Source: D. Biello, "How to Use Solar Energy at Night," *Scientific American*, February 18, 2009.

There are a number of proposed and planned CSP plants incorporating TES. In 2011, construction began in Arizona on a 250-MW parabolic trough plant with six hours of storage.[419] CSP plants are currently incentivized by a 30% federal investment tax credit (ITC), currently scheduled to expire in 2016, and the Department of Energy's Loan Guarantee Program amended by the American Recovery and Reinvestment Act of 2009. The first loan guarantee of $1.45 billion was used to finance construction and start-up of the Arizona project.[420]

Technology

Description

With thermal storage, the heat from the solar field is stored prior to reaching the power generation turbine. Storage media being used or considered include molten salt, steam accumulators (for short-term storage only), solid ceramic particles, high-temperature phase-change materials, graphite, and high-temperature concrete.[421] Molten salt is the most commonly used storage medium and is used in the two-tank storage system shown in **Figure 36**. In this approach, hot heat-transfer fluid (HTF), typically a synthetic oil, from the solar field flows through heat exchangers to charge molten salt in the "hot" storage tank. This salt starts out as lower-temperature material in the "cold" storage tank, having come from the power block after being used to generate steam for the steam turbine/generator. When the energy in the hot storage tank is

---

[419] Excluding pumped hydro plants, this is larger than any other single storage facility in the United States.

[420] U.S. Department of Energy, "DOE Finalizes $1.45 Billion Loan Guarantee for One of the World's Largest Solar Generation Plants," press release, December 21, 2010.

[421] A. Gil, et al., "State of the Art on High Temperature Thermal Energy Storage for Power Generation. Part 1—Concepts, Materials and Modellization," *Renewable and Sustainable Energy Review*, No. 14, pp. 31-55, 2010; D. Kearney et al., "Engineering Aspects of a Molten Salt Heat Transfer Fluid in a Trough Solar Field," *Energy*, No. 29, pp. 861-870, 2004; U. Herrman and D. Kearney, "Survey of Thermal Energy Storage for Parabolic Trough Power Plants," Journal of Solar Energy Engineering, No. 124, pp. 145-152, 2002.

---

needed, the system simply operates in reverse to reheat the solar HTF. The hot HTF then releases its thermal energy through a set of heat exchangers to generate steam to spin the turbine and generator in the power plant. It is an "indirect" system because it uses a storage medium (salt) that is different from the HTF (oil) circulated in the solar field.

Other options include direct storage. This includes use of molten salt as the HTF, such as is used in the Gemasolar plant. Another option is the direct storage of steam, which is being used commercially in Spain.[422] Steam storage is typically limited to less than 1 hour of generation due to the high cost of pressurized vessels for larger steam volumes and storage capacities.

Figure 36. Schematic of an Indirect Two-Tank TES System

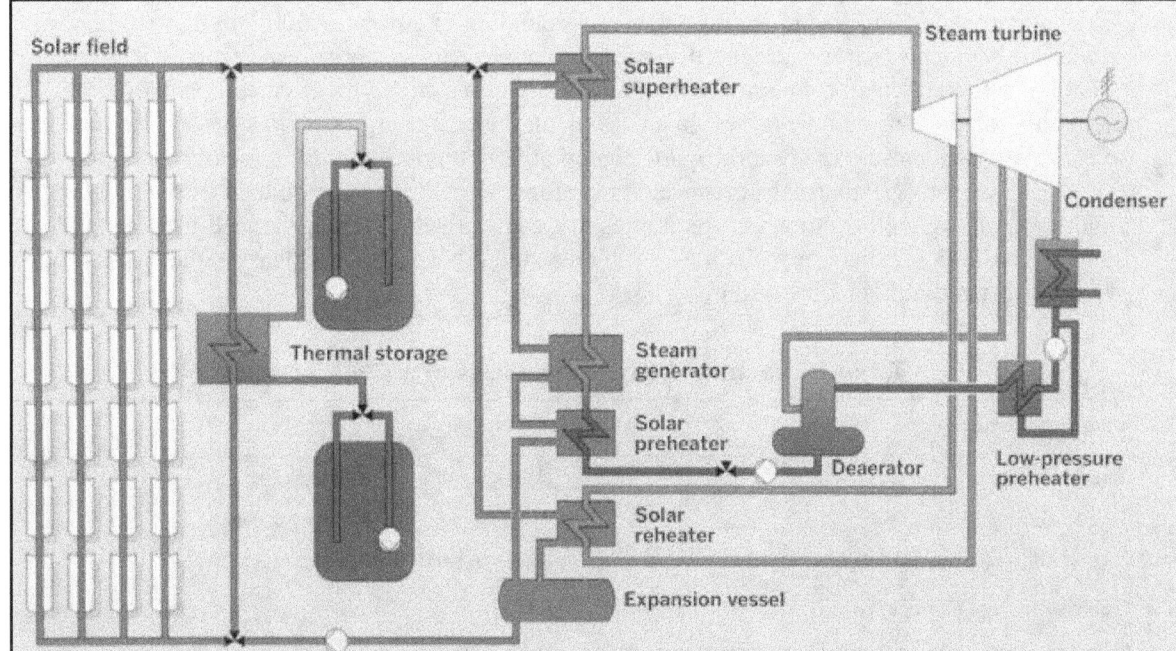

Source: A. Gil, et al., 2010.

Performance

TES can greatly improve the performance of CSP plants. Low levels of storage—30 minutes to 1 hour of full-load storage—can ease the impact of transients such as clouds. However, the most significant attribute of thermal storage is that it can significantly increase the energy and capacity value of CSP compared to equivalent systems without storage. In addition, CSP systems with storage can provide load following and ancillary services acting as an enabling technology for variable generation sources.[423]

Trough plants such as Andasol 1 have been designed for 7.5 hours of full-load storage, which allows operation well into the evening when peak demand can occur and energy costs are high.

---

[422] Abengoa Solar, *Power Tower Technology Plants*, marketing brochure, 2011. http://www.abengoasolar.com/corp/export/sites/solar/resources/pdf/en/PS10.pdf

[423] R. Sioshansi and P. Denholm, 2010.

Power towers, with their higher operating temperatures, can charge and store molten salt more efficiently and less expensively.[424] The 17-MW Gemasolar power tower being developed in Spain is designed for an operation of 6500 hours per year—or a 74% capacity factor. In Arizona, a power purchase agreement has been signed for the Solana project, a 250-MW net trough plant with 6 hours of molten-salt thermal storage, which yields a 40% capacity factor.

CSP trough and tower plants generate electricity at utility-scale levels, currently ranging from 10 MW to 50 MW or more. The single power block—the portion of the plant where power is generated—for each of these plants is one reason that TES can readily be incorporated. For dish/engine CSP systems, in contrast, the potential for TES is currently much less because of their more modular design and much lower (25 kW or less) generating capacities.

A key advantage of TES over electricity storage technologies is very high round-trip efficiency, because it is much easier to store thermal energy than electricity energy. Electricity is a high "quality" source of energy, so transforming electricity into a stored medium and back to electricity incurs considerable losses. In a CSP plant, thermal energy is stored before conversion to electricity. As a result, the round-trip efficiency of CSP thermal storage typically exceeds 90%.[425] However, CSP thermal storage can only store thermal energy produced from the solar field, as opposed to other storage technologies that can store electricity produced from any source. **Table 13** summarizes the storage efficiency and other basic technical characteristics of a typical two-tank indirect TES system.

Table 13. Technical Characteristics of a CSP System

Characteristic	Description
Energy density	~ 155 kJ/kg of molten salt[a]
Power density	Determined by heat-exchanger design
Size of typical installations	Power: 50 MW, Energy: 350 MWh
Thermal storage efficiency	>90%
Response time (to bring thermal energy from storage to the steam generator)	~ 10 minutes
Decay (of stored energy)	~1%/day
Lifetime	30 years (same as lifetime of power plant)

Source: U. Herrman and D. Kearney, "Survey of Thermal Energy Storage for Parabolic Trough Power Plants," Journal of Solar Energy Engineering, No. 124, 2002, pp.145-152.

a.   This value (for a trough) is roughly one-third of that for current power towers.

---

[424] A. Gil, et al. 2010.

[425] Performance measurements from the Solar Two plant estimated a daily loss of about 3.4% of thermal energy, representing a >96% storage efficiency for a plant cycled daily. The report estimates that that a commercial plant would experience annual storage losses of less than 1% due to the higher volume to surface area ratio of the storage tanks and the fact that most stored energy is used the same day. J. E. Pacheco, Editor, Final Test and Evaluation Results from the Solar Two Project, SAND2002-0120, Sandia National Laboratories, January 2002.

Cost

One recent estimate for the cost of two-tank indirect molten salt TES in a trough system is approximately $240/kWh of electricity output.[426] For a two-tank direct molten salt power tower, the cost estimate is $72/kWh. The higher temperature and other aspects of towers result in the use of about one-third the amount of molten salt compared to troughs. While these costs are lower than those for most other storage technologies, a direct comparison is of limited value since TES must be tied to a single solar plant. The cost of CSP plants is typically compared to a conventional alternative, such as a combined-cycle natural gas plant, while potentially considering the added benefits of zero-emissions and fuel security. One estimate for the levelized cost of electricity for a trough plant is about 19 ¢/kWh (2009 dollars).

In 2009, the DOE set a goal to establish CSP technology as competitive with conventional intermediate-load generation technologies by 2020.[427] The targeted cost of energy was approximately 8-12 ¢/kWh, depending on market conditions.[428] In 2011, the DOE officially unveiled the SunShot Initiative, an aggressive R&D plan to make large-scale solar energy systems cost competitive without subsidies by 2020. The SunShot Initiative takes a systems-level approach to revolutionary, disruptive (as opposed to incremental) technological advancements in the field of solar energy. The goal of the SunShot Initiative is reaching cost parity with baseload energy rates, which would pave the way for rapid and large-scale adoption of solar electricity across the United States.

Cost reduction of current CSP technologies will result, in part, from economies of scale and from learning due to increased production and deployment of CSP systems. But a significant fraction of cost reduction will also need to come from improvements in performance and costs of the solar field, as well as TES. Once of the key drivers in the cost of TES is the relatively small temperature difference between the cold and hot fluid in the storage system which increases the volume of storage medium required.[429] A variety of options for cost reduction and technology improvements exist as discussed in the following section.

In general, very little operations and maintenance (O&M) should be required for the TES systems. However, reliability of certain components such as the molten salt pumps may affect the long-term O&M for TES. The Andasol plants in Spain started operating in 2009. O&M costs for TES are currently being determined at these plants; it will require several years of operation before they can be accurately assessed.

---

[426] C. Turchi, M. Mehos, C.K. Ho, and G. J. Kolb, "Current and Future Costs for Parabolic Trough and Power Tower Systems in the U.S. Market," Conference Paper, NREL/CP-5500-49303, National Renewable Energy Laboratory, October 2010.

[427] Ibid.; For additional information about cost improvement potential, see Sargent & Lundy LLC, "Assessment of Parabolic Trough and Power Tower Solar Technology Cost and Performance Forecasts," subcontractor report, NREL/SR-550-34440, National Renewable Energy Laboratory, October 2003.

[428] C. Kutscher, et al., *Line-Focus Solar Power Plant Cost Reduction Plan (Milestone Report,* No. TP-5500-48175, National Renewable Energy Laboratory, 2010.

[429] C. Turchi, "Parabolic Trough Reference Plant for Cost Modeling with the Solar Advisor Model (SAM)," NREL/TP-550-47605, National Renewable Energy Laboratory, July 2010.

Research and Development

A major focus on TES research and development is developing new TES materials and storage methods. As noted above, the DOE has prepared roadmaps for line-focus and power tower technologies, identifying and prioritizing Technology Improvement Opportunities (TIOs) to help meet program cost goals.[430] The TIOs with greatest importance are the following:

- High-temperature heat transfer fluids (HTFs) with freeze protection,
- Low-cost storage fluids and media,
- High-energy-density phase-change materials (PCMs) and storage fluid, and
- High-energy-density thermochemical TES.

Alternative TES materials, including phase-change materials, graphite, and concrete each offer some combination of greater stored energy density and lower cost. These materials are currently not in use but are being developed as advanced storage materials. Ceramic particles, graphite, and concrete are typically used in a thermocline system because these materials store thermal energy as solids and cannot be transferred easily like a molten salt or other fluid. A thermocline is a storage system in which the storage medium (solid) is stationary in a packed bed or monolithic structure.[431] A heat transfer fluid flows through the bed or structure and transfers thermal energy to and from the solid storage material as the system is charged and discharged. These materials are generally low cost, but control of this type of storage system is difficult and can lead to losses in plant efficiency.

Research efforts in the United States on advanced CSP thermal storage materials include activities at NREL, Sandia, and through the Funding Opportunity Announcement (FOA) awards granted to 18 recipients by the DOE in 2007 and 2008.[432] **Table 14** shows a list of the FOAs, being performed by a combination of universities and companies, categorized by TIO. This list indicates that there are a large number of pathways that may improve the performance of TES technologies.

---

[430] J.A. Gary, et al., "Development of a Power Tower Technology Roadmap For DOE," SAND2010-5279C, Sandia National Laboratories, 2010.

[431] G. J. Kolb, "Evaluation of Annual Performance of 2-Tank and Thermocline Thermal Storage for Trough Plants," *Journal of Solar Energy Engineering*, Vol. 133, No. 3, August 2011.

[432] J. Stekli, "DOE CSP R&D: Storage Award Overview," slide presentation, U.S. Department of Energy, April 28, 2010. http://www1.eere.energy.gov/solar/pdfs/storage_award_progress_seia.pdf

Table 14. U.S. Department of Energy FOA Projects

Technology Improvement Opportunities	Institution (Research Area)
Increased Temperature Differential (ΔT)	Abengoa Solar (Molten-salt HTF)
	Skyfuel (High-temperature linear Fresnel)
	Solar Millennium (High-temperature collector)
	Halotechnics (Eutectic salt formulations)
Low-cost media	Acciona (Sensible, direct TES)
	University of Arkansas (High-temperature concrete)
	U.S. Solar Holdings (Thermocline demonstration)
	U.S. Solar Holdings (Sand-shifter demonstration)
High-energy-density PCMs	Lehigh University (Encapsulated PCMs)
	Terrafore (Off-eutectic formulations)
	University of Connecticut (PCM thermosyphons)
	Abengoa Solar (Cascading PCMs)
	Infinia (PCMs for dish/Stirling)
	Acciona (PCMs with agitation)
Low-cost storage fluids/media	University of Alabama (Low-melting-point, low-cost salts)
High-energy-density storage fluid	Texas A&M University (Storage nanofluids)
High-energy-density thermochemical TES	General Atomics (High-temperature thermochemical TES)

Source: J. Stekli, 2010.

Deployment of advanced TES could contribute significantly to cost reduction for CSP/TES systems. **Table 15** shows the results of an analysis of potential reduction in levelized cost of energy (LCOE) related to technology improvement opportunities in thermal energy storage.

Table 15. Potential Cost Reductions for CSP/TES Systems

Technology Improvement Opportunity	LCOE Reduction (¢/kWh)	LCOE Reduction (%)
Increased ΔT: 500°C w/o freeze protection	2.0	11
Increased ΔT: 500°C with freeze protection	1.7	9
Low-cost media: thermocline	1.1	6
Low-cost media: sand	1.1	6
High-energy-density phase-change materials	1.5	8
Low-cost storage fluids/media	0.8	4
High-energy-density storage fluid	1.5	8
High-energy-density thermochemical TES	3.4	18

Source: J. Stekli, 2010.

Deployment Challenges

The primary barrier to deployment of TES with CSP is more associated with the solar plant than the thermal storage component. TES is only an option where CSP is deployable—areas with high direct normal radiation—such as the southwestern United States. Although this region is close to some major load centers, especially Southern California, it would require large-scale development of new transmission to provide significant electricity supply to eastern states.

The use of molten salts adds little additional environmental impact or risk beyond a standard CSP plant. The footprint of the tanks is small compared to the footprint of the solar field. A failure of the tank would result in a spill that would be contained by a wall or berm. The salt is a solid at ambient temperatures, so the liquid salt would quickly freeze into a non-hazardous solid.

Material requirements for TES are not expected to be a significant constraint, although large-scale deployment both in the United States and internationally, combined with usage at the rate associated with current two-tank systems, could require development of new sources. The molten salt requirement for a 100 MW trough plant with six hours of storage using a two-tank system is about 57,000 metric tons, while a tower system would require about a third of this amount.[433] Deployment of thermoclines would also substantially reduce this requirement. Much of the world's nitrate salts are derived from deposits in the Atacama region of Chile. Proven reserves are 29.4 million metric tons, although this figure is based on exploration of only 16% of total reserves.[434] An alternative source of nitrite salts could include synthetic production via the Haber-Bosch process, used worldwide for fertilizer production, while entirely new TES materials could also provide alternatives to molten salts.

Conclusions

Thermal energy storage is a commercially available option for large-scale deployment, featuring higher efficiency than most other electricity storage technology, although tied to a single application with significant geographical constraints. Deployment of TES is dependent on cost-competitiveness of CSP, and significant cost barriers remain to achieving large-scale deployment of TES with CSP. However, there are multiple pathways to improve CSP plant efficiency and reduce the delivered cost of energy via higher-temperature parabolic troughs and power tower technologies. Higher-temperatures also increase the storage capacity of the TES system. The currently operating plants are expected to demonstrate good reliability and performance. Assuming this positive outcome, and success in efforts to increase technical performance, CSP/TES will continue be a viable option to decrease the variability and increase the dispatchability of the solar resource in areas where CSP is economically viable.

---

[433] J.J. Burkhardt III, G. A. Heath, and C.S. Turchi, "Life Cycle Assessment Of A Parabolic Trough Concentrating Solar Power Plant And The Impacts Of Key Design Alternatives," *Environmental Science & Technology*, Vol. 45, No. 6., 2011, pp. 2457-2464.

[434] Sociedad Quimica y Minera de Chile S.A. (SQM)., Form 20-F for U.S. Securities & Exchange Commission, June 30, 2010. http://www.sqm.com/PDF/Investors/20F/SQM-20F_2009-(30%20June%202009)-EN.pdf

Chapter 13: Superconducting Magnetic Energy Storage

Overview

Superconducting magnetic energy storage (SMES) devices store energy in the form of a magnetic field. Using superconducting wires allows SMES units to achieve very high efficiencies. Superconducting materials are characterized by the temperature required to achieve superconductivity: low-temperature or high-temperature. Low-temperature superconductors, which require expensive liquid helium, have been used in most SMES demonstrations. Recent SMES research and development have focused on developing high-temperature SMES devices, which could use less costly liquid nitrogen and therefore, could have lower overall system costs that the low-temperature SMES devices demonstrated to date.[435] While low-temperature SMES devices have been demonstrated, their costs have been greater than those of proposed high-temperature systems.[436]

There have been several SMES demonstration projects for quick-response and very short-term capacity applications, primarily for electric grid power quality.[437] Commercial SMES units typically have relatively large power capacities (on the order of several MW), but short discharge times on the order of 1 second, because the high cost of superconducting coils and cryocoolers limits amount of energy stored. One example is a set of 3 MW/0.8 kWh distributed SMES units built in 2000 by the American Semiconductor company for Wisconsin Public Service Corporation to enhance voltage stability in northern Wisconsin.[438] Other utilities have purchased similar units to provide voltage stability and power quality, and to defer the construction of new transmission lines. Since the distributed SMES units are trailer-mounted, they can be optimally sited at locations on the grid where they are most needed. Distributed SMES systems have also been installed in industrial applications, primarily to eliminate voltage sags in manufacturing power supplies.[439] The cost of SMES is currently very high, however; significant reductions are needed to show an economic advantage over alternatives including batteries, capacitors, and power electronics alternatives.

---

[435] A.M.Wolsky, "The Status and Prospects for Flywheels and SMES that Incorporate HTS," *Physica C*, Vol. 372, 2002, pp.1495-1499; J.-F. Fagnard, et al., "Use of a High-Temperature Superconducting Coil for Magnetic Energy Storage," *Journal of Physics*, Conference Series 43, 2006, pp. 829-832; J.H. Choi et al., "Basic Insulating Design of a Magnet Coil and a Bobbin for a Conduction Cooled HTS SMES System," *Physica C*, Vol. 463, 2007, pp. 1252-1256; K.C. Seong, "Development of a 600 kJ HTS SMES," *Physica C*, Vol. 468, 2008, pp. 2091-2095.

[436] H.C. Freyhardt, "YBaCuO and REBaCuO HTS for Applications," *International Journal of Applied Ceramic Technology*, Vol. 4, No. 3, 2007, pp. 203-216; EPRI/DOE, 2003.

[437] D. Sutanto and K. Cheng, "Superconducting Magnetic Energy Storage Systems for Power System Applications," proceedings of the International Conference on Applied Superconductivity and Electromagnetic Devices, Institute of Electrical and Electronics Engineers, 2009, pp. 377-380; EPRI/DOE, 2003.

[438] J.B. Howe, Distributed SMES: A New Technology Supporting Active Grid Management, *Modern Power Systems*, January 22, 2001.

[439] Chubu Electric Power Company, Inc., "Field Testing of Superconducting Magnetic Energy Storage System (SMES)," press release, February 21, 2003, http://www.chuden.co.jp/english/corporate/press2002/0221_1.html. This is a 10MW unit at the Sharp Corp. Kameyama Plant.

Technology

Description

SMES devices are large superconducting electromagnets that store energy in a magnetic field generated by electric current flowing through superconducting magnetic wire. The wire is typically coiled in loops to form a solenoid. Because superconducting material has no electrical resistance, very large amounts of current can be sent through these wires, up to a factor or 100-500 greater than equivalently sized copper wire.[440] This enables very strong magnetic fields (measured in tens to hundreds of Teslas), and much stronger energy densities than conventional electromagnets.[441] The amount of energy that can be stored in the resulting magnetic field quadruples for each doubling of current, but also depends on the coil geometry and the magnetic permeability of the material inside and surrounding the coil.

SMES systems typically consist of four main components: the superconducting coil, the cryogenic cooling system, the power conditioning system, and the control unit. Superconducting materials are characterized by the temperature required to achieve superconducting characteristics. Low temperature superconductors (LTS), like niobium-titanium (NbTi), have been used in most commercial SMES applications.[442] The electromagnetic coils must be cooled to 4.5°K (-269°C) to become superconducting, which is typically done with a cryocooler system that uses liquid helium as the working fluid.[443] Liquid helium, with a boiling point of 4.2°K, is the only element that is not solid at this low temperature. High temperature semiconductors (HTS), such as $(Bi,Pb)_2Sr_2Ca_2Cu_3O_x$ (BSCCO) and $YBa_2Cu_3O_x$ (YBCO), are superconductive at the boiling point of liquid nitrogen (77 °K).[444] The SMES coil is typically made out of LTS materials, typically NbTi.[445] HTS prototypes have been demonstrated using BSCCO, and there is ongoing work to demonstrate less expensive YBCO coils.[446] LTS SMES devices require the coil to be cooled to about 4.5°K, whereas HTS SMES devices require cooling to only about 77°K.[447] For this reason, LTS cryocoolers use liquid helium as the working fluid and HTS cryocoolers use liquid nitrogen which is far less expensive. In both systems, the cryocooler must be located outside the area with a strong magnetic field, with a thermally conductive link between the cryocooler and coil.[448] The power electronics unit converts alternating current (AC) electricity

---

[440] The current carrying capacity of a superconductor is about 300,000 Amps/cm², whereas the current carrying capacity of an uncooled copper wire is about 600 Amps/cm², and the capacity of an actively cooled copper wire is about 3,000 Amps/cm²; EPRI, 2003.

[441] Tesla is a unit used to characterize the strength of a magnetic field. The Earth's magnetic field is about 10^{-5} Teslas near the surface. A common refrigerator magnet is about 10^{-3} Teslas.

[442] EPRI/DOE, 2003.

[443] Temperature units are in degrees Kelvin (°K). The temperature difference measured by one degree K is equal to one degree C, but the Kelvin scale is zeroed at absolute zero (0 °K equals about -273 °C) whereas the Celsius scale is zeroed at the freezing point of water (0 °C equals about 273 °K).

[444] D. Larbalestier et al., "High-Tc Superconducting Materials for Electric Power Applications," *Nature*, No. 414, 2001, pp. 368-377; H.C. Freyhardt, "YBaCuO and REBaCuO HTS for Applications," International Journal of Applied Ceramic Technology, No. 4, 2007, pp. 203-216.

[445] EPRI/DOE, 2003.

[446] J.-F. Fagnard, et al. 2006; K.C. Seong et al., 2008; D. Larbalestier et al., 2001; H.C. Freyhardt, 2007.

[447] K.C. Seong et al., 2008.

[448] Y.S. Choi et al., "Cryocooled Cooling System for Superconducting Magnet," *Cryocoolers 15*, edited by S.D. Miller and R.G. Ross, Jr., International Cryocooler Conference, Inc., Boulder, CO, 2009.

---

from the electric grid to direct current (DC) electricity to be stored when charging, reversing this process during discharge.

Performance

Typical SMES performance is characterized by short bursts (about 1 second) of power that are capable of responding very rapidly (less than 0.5 milliseconds) from fault detection, well within one cycle of power grid frequency.[449] SMES devices are designed to provide tens of thousands of charge/discharge cycles with little or no performance degradation and have long design lifetimes on the order of 20 years operating continuously.[450] **Table 16** summarizes key SMES operating characteristics.

Table 16. SMES Operating Parameters

Operating parameter	Value
SMES capacity density	160 kW/m²
SMES energy density	0.04 kWh/m²
Response rate	< 1 cycle (0.017 seconds)
Instantaneous system efficiency	96%-98%
Round trip efficiency	Up to 95%; Highly dependent on operating characteristics
Standby energy losses	1%/hr
Design lifetime	20 years

Source: E. Drury, National Renewable Energy Laboratory, 2009. Compiled from EPRI/DOE, 2003; A.M. Wolsky, 2002; H.C. Freyhardt, 2007.

The roundtrip efficiency of an SMES device is based on the instantaneous system efficiency (including energy conversion losses and standby energy losses) as well as the energy required to maintain cool temperatures. Instantaneous efficiency has been estimated to range from 96% to 98%, before parasitic and power conversion losses.[451] Cryocooler power loads are typically about 1% of nameplate capacity for LTS SMES units.[452] Cryocooler loads could decrease by an order of magnitude or more for HTS SMES units.[453] SMES roundtrip efficiencies are also dependent on how frequently they are charged and discharged. For example, if SMES systems are charged and discharged every hour, roundtrip efficiency is greater than 95%. If SMES systems discharge once per day, roundtrip efficiency drops to 73%. However, since commercial SMES applications provide a capacity resource, and not an energy resource, roundtrip efficiency is far less important than other factors like capital and operating costs.

---

[449] J.B. Howe, 2001.

[450] EPRI/DOE, 2003.

[451] EPRI/DOE, 2003.

[452] A.M. Wolsky, 2002.

[453] The power required to remove heat from the superconducting coil increases with decreasing temperature. At 4.5 °K (LTS system), about 200 to 1,000 watts of electric power are required to remove one watt of heat seep into the coil whereas at 77 °K (HTS system), about 20 watts are required. EPRI/DOE, 2003; K.B. Wilson and D.R. Gedeon, "Status of Pulse Tube Cryocooler Development at Sunpower, Inc.," proceedings of the International Cryocooler Conference, Cambridge, MA, March 29-April 1, 2004, Kluwer Academic/Plenum Press, NY, 2004.

Cost

SMES costs are driven primarily by the cost of the superconductive coil and cryocooler, limiting the amount of energy that can be stored economically. Power electronics costs are based on the capacity of the device and range from $195-$325/kW (2009 dollars).[454] The superconductive coil and cryocooler costs scale with the amount of energy to be stored by the SMES device, and range from $395,000-$740,000/kWh. **Figure 37** integrates the capacity-based and energy-based costs into device cost estimates for a design range of 1 to 30 seconds of energy storage. The cost of a commercially produced SMES unit (10MW/2.8kWh, 1 second discharge) is estimated at $215/kW[455]-$285/kW,[456] and is within the lower range of projected costs. Since very few SMES devices have been built commercially, these costs represent demonstration project costs. It is likely that future costs could benefit from organizational learning-based cost reductions and by achieving economies of scale.

Figure 37. Potential SMES Cost Ranges Based on Component Costs

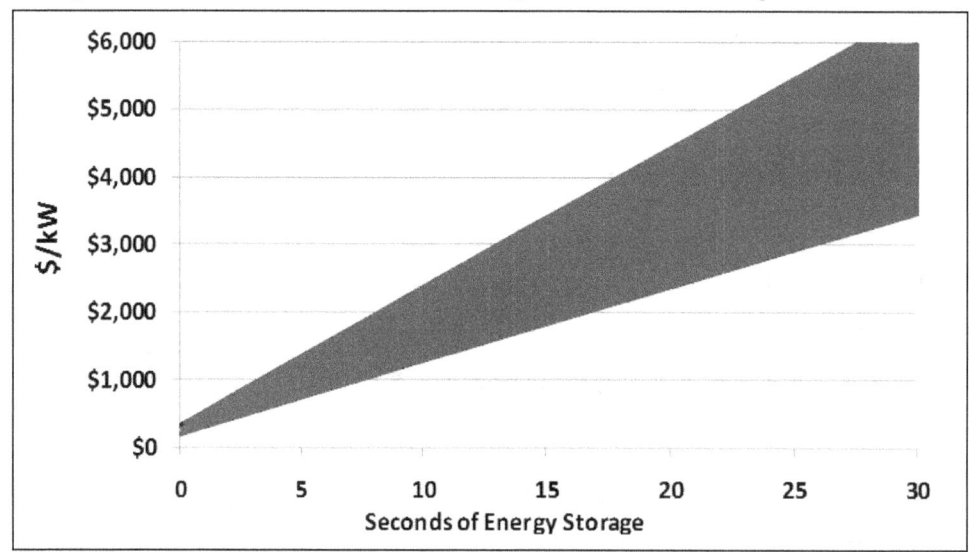

Source: E. Drury, National Renewable Energy Laboratory, 2009. Costs are 2009 dollars, excluding O&M.

SMES costs increase significantly with increasing energy storage. A $1,000/kW SMES device is likely to have the stored energy capacity to discharge for 3.5-7.5 seconds, and a $2,000/kW device has a stored energy range of about 8-16 seconds. Unless energy-based costs for SMES devices achieve significant cost improvements, SMES will likely only compete in power quality markets. Both the superconductor and cryocooler costs would have to achieve about an order of magnitude of cost reductions to compete with flywheels and batteries in regulation and reserves markets, and they would have to achieve several orders of magnitude cost reductions to compete with compressed air energy storage and pumped storage hydro in diurnal storage applications.

---

[454] V. Karasik, et al., "SMES for Power Utility Applications: A Review of Technical and Cost Considerations," *IEEE Transactions on Applied Superconductivity*, Vol. 9, 1999, pp. 541-546.

[455] L. Borgard, "Grid Voltage Support at Your Fingertips," *Transmission and Distribution World*, October 1, 1999.

[456] EPRI/DOE, 2003.

O&M costs are not included in **Figure 37**. Fixed O&M costs have been estimated at $16-$26/kW-yr.[457] Variable O&M costs have been estimated at about $11-$14/kW-yr,[458] of which about $5/kW-yr is based on the cost of electricity for cooling an LTS system.[459] A large fraction of variable O&M costs are dependent on the electricity costs associated with running cryocoolers.

Research and Development

SMES research and development is focused primarily on developing and demonstrating HTS coil materials, as well as improving cryocooler design and performance. Several BSCCO HTS storage devices have been built and demonstrated in laboratories.[460] However, no YBCO devices have been demonstrated, although YBCO has the potential to significantly reduce HTS costs if fabrication challenges can be overcome.[461] Currently, HTS systems remain more expensive than LTS systems, which remain the commercial standard. Cryocoolers are also an active area of research, much of which has focused on improving designs to reduce initial capital costs, and improving performance to reduce cryocooler power loads, and operating costs.[462]

Deployment Challenges

SMES systems do not use fuel or water, do not contain hazardous chemicals, are emissions free, and have a relatively small footprint. Additionally, SMES units are frequently designed to be mobile.[463] One challenge in siting a SMES resource is limiting human exposure to strong magnetic fields.[464] Magnetic field exposure can be managed through magnetic shielding (using passive material shielding, or active shielding with a compensating magnetic coil) or by siting the SMES resource on enough land to limit human exposure.

It is not likely that material constraints will limit SMES deployment. LTS superconductors require niobium and titanium. Niobium is rare, but annual niobium production is tens of thousands of tons, which is more than sufficient for potential SMES applications. Titanium is a fairly common earth element mined at a rate of millions of tons annually. The elements required for the HTS material YBCO are not rare, and will not limit deployment.

Conclusions

SMES systems are likely to be limited to power quality applications in utility and industrial markets. SMES costs are primarily driven by the cost of superconducting coils and cryocoolers. These costs would have to decrease by at least an order of magnitude for SMES to compete in

---

[457] EPRI/DOE, 2003.

[458] EPRI/DOE, 2003.

[459] A.M. Wolsky, 2002. Cost estimate based on wholesale electricity rates for Wisconsin Public Service Corporation.

[460] A.M. Wolsky, 2002; J.-F. Fagnard et al., 2006; Y.S. Choi et al., 2007; H.C. Freyhardt, 2007; K.C. Seong et al., 2008.

[461] D. Larbalestier et al., 2001; H.C. Freyhardt, 2007.

[462] K.B. Wilson and D.R. Gedeon; 2004; K.C. Seong et al., 2008; Y.S. Choi et al., 2009.

[463] EPRI/DOE, 2003.

[464] C. Polk, R.W. Boom, and Y.M. Eyssa, "Superconductive Magnetic Energy Storage (SMES) External Fields and Safety Considerations," *IEEE Transactions on Magnetics*, No. 28, 1992, pp. 478-481.

frequency regulation and reserve markets, and by several orders of magnitude to compete in diurnal storage markets.

SMES research and development are currently focused on developing and demonstrating high temperature superconductor materials for the electromagnetic coil, as well as improving cryocooler design and performance. HTS units have historically used BSCCO materials, and there is also the potential to demonstrate the use of less costly HTS materials like YBCO. Currently, HTS magnetic energy storage systems are more expensive than low temperature superconducting LTS systems, which remain the commercial standard. There is considerable effort on developing and demonstrating new HTS materials across several fields. SMES costs will benefit from future advancements driven by developments in related fields. However, HTS costs will have to decrease substantially for SMES to compete in energy-intensive storage markets.

Appendix. Table of Acronyms

AC - alternating current

AFC - alkaline fuel cell

ARPA-E - Advanced Research Projects Agency – Energy

ARRA - American Recovery and Reinvestment Act

BSCCO - bismuth strontium calcium copper oxide (type of superconductor)

CAES - compressed air energy storage

CAISO - California Independent System Operator

CEC - California Energy Commission

CER - charging electricity ratio

CSP - concentrating solar power

CT - combustion turbine

DC - direct current

DMFC- direct methanol fuel cell

DoD - depth of discharge

DOE - Department of Energy

EC - electrochemical capacitors

EPRI - Electric Power Research Institute

ERCOT - Electric Reliability Council of Texas

EV - electric vehicle

FCEV - fuel cell electric vehicle

FERC - Federal Energy Regulatory Commission

FOA - funding opportunity announcement

GW - gigawatts

HEV - hybrid electric vehicle

HHV - higher heating value

HTF - heat-transfer fluid

ISO - Independent System Operator

ITC - investment tax credit

KOH - potassium hydroxide

kW - kilowatt

kWh - kilowatt hour

Li-Ion - lithium-ion

LTS - low temperature superconductor

MCFC - molten carbonate fuel cell

MWh - megawatt hour

NaS - sodium sulfur

NERC - North American Electric Reliability Corporation

NiMH - nickel metal hydride

NOx - nitrogen oxides

NREL - National Renewable Energy Laboratory

NYSERDA - New York State Energy Research and Development Authority

O&M - operations and maintenance

OCC - overnight construction cost

PAFC - phosphoric acid

PEM - polymer electrolyte membrane

PEMFC - proton exchange membrane fuel cells

PHEV - plug-in hybrid electric vehicles

PHS - pumped hydro storage

PSI - pounds per square inch

R&D - research and development

RD&D - research, development, and deployment

RTO - regional transmission organization

SLI - starting, lighting, and ignition

SMES - superconducting magnetic energy storage

SOFC - solid oxide fuel cell

T&D - transmission and distribution

TES - Thermal energy storage

TIO - technology improvement opportunities

TOU - time of use

TWh - terrawatt hour

UPS - uninterruptible power supply

V2G - hehicle-to-grid

VG - variable generation

VMT - vehicle mile traveled

VOC - volatile organic compound

WWSIS - Western Wind and Solar Integration Study

YBCO - yttrium barium copper oxide (type of superconductor)

Author Contact Information

Paul W. Parfomak
Specialist in Energy and Infrastructure Policy
pparfomak@crs.loc.gov, 7-0030

www.ingramcontent.com/pod-product-compliance
Lightning Source LLC
Chambersburg PA
CBHW081453170526
45166CB00008B/2419